OBJECTIVE ANIMAL PHYSIOLOGY

FOR
JRF, SRF, ARS, NET, CIVIL SERVICES & OTHER COMPETITIVE EXAMINATIONS

By

Dr. Arpita Mohapatra
Ph.D. (Animal Physiology)
Scientist, ICAR-CSWRI
Avikanagar, Rajasthan

Dr. Shakti Ranjan Panigrahy
B.V. Sc & A.H, MBA (AGRIBUSINESS)
Assistant Professor
IABMI, AAU, Anand

Competition Tutor
An imprint of Scientific Publishers (India)

Published by:
Competition Tutor
5-B, New Pali Road, P.O. Box 91
Jodhpur 342 001 (India)

E-mail: competitiontutor@gmail.com
Website: www.competitiontutor.co.in

An imprint of Scientific Publishers (India)

ISBN: 978-93-86652-46-1

Printed in India

:: Dedication ::

This book is dedicated to our loving daughter "**Shatabdi**" and cute son "**Aayushman**". Thanks God for giving us a successful married life and two living angels who make our life complete. We are extremely thankful to our parents without whose support it is completely impossible to fulfill this endeavor.

Preface

Dedication, Determination and Discipline are the 3Ds for success. Nothing is more informative than a textbook. But after reading a book, to become more confident about your preparation before an examination, we need to access our self. When I was preparing for Agricultural Research Service (ARS) I followed textbooks. Which had given me background knowledge of Animal Physiology. But before the examination it is not at all possible to revise the whole book again for boosting my self-confidence. That time I desperately need a handy and accessible guide who can speed up my revision by asking questions from textbook. This give me an idea to write an objective book in discipline of Animal Physiology.

The book, **"OBJECTIVE ANIMAL PHYSIOLOGY"** is a complete package for the aspirants for ICAR (NET), JRF, SRF, ARS and Civil Service Examination in the subject category of Animal Physiology. This book constitutes more than one thousand four hundred questions that pool the whole animal physiology in one platform. It covers all the sub disciplines like behavior, climatology, growth, lactation, circulatory system, endocrinology, reproduction, digestion, excretion, respiration etc. This book definitely enhance the confidence of students by building more clarity about the subject. This is our initial approach and we believe that with increase in our experience this book will encompass more and more questions of Animal Physiology and emerge out as an encyclopedia in this subject. This book has been divided in to nine chapters so that a reader can proceed step by step. Examination oriented questions and their accuracy enhances the quality of the book. This book will act as a mentor for the students who want to find themselves in eligibility list for teaching and research.

– Arpita Mohapatra

– Shakti Ranjan Panigrahy

Contents

• Chapter 1 •

Behaviour, Growth and Climatology

1. **The study of animal behaviour in natural environment is called :**
(A) Etiology (B) Psychology (C) Ethology (D) Parapsychology

2. **The example of associative learning is :**
(A) Classical conditioning (B) Operant conditioning
(C) Pavlovian conditioning (D) All of the above

3. **Increased response to an increase in light intensity is called as :**
(A) Positive phototaxis (B) Kinesis
(C) Negative phototaxis (D) Luminis

4. **A "Skinner box" is used for experiments in :**
(A) Classical conditioning (B) Operant conditioning
(C) Learning preparedness (D) All of these

5. **Which of the following involves trial and error learning :**
(A) Habituation (B) Sensitization
(C) Classical conditioning (D) Operant conditioning

6. **Chemical signals between individuals of same species :**
(A) Hormone (B) Cytokines
(C) Pheromones (D) All of the above

7. **Circadian rhythms are based on approximately :**
(A) 1- hour period (B) 12- hour period
(C) 24- hour period (D) 72- hour period

8. **The biological clock of mammals is located at :**
(A) Suprachiasmatic nucleus of hypothalamus
(B) Suprachiasmatic nucleus of optic chiasma
(C) Suprachiasmatic nucleus of pineal gland
(D) Gonads

9. **Dog salivating at the sound of a can opener is an example of :**
(A) Classical conditioning (B) Operant conditioning
(C) Fixed action pattern (D) Imprinting

10. **Squirrels in city park ignoring sounds of moving vehicles is an example of :**
(A) Classical conditioning (B) Operant conditioning
(C) Habituation (D) Sensitization

11. **A rat in a box learns to associate pressing of lever with obtaining food :**
(A) Classical conditioning (B) Fixed action pattern
(C) Habituation (D) Operant conditioning

12. **Courtship behaviour in animals involves all the following except :**
(A) Visual cues
(B) Acoustic signal
(C) Chemical signal
(D) All the above are involved in courtship behaviour

13. A type of learning in which an animal, at a critical time of its life, forms a social attachment of another object is called :

(A) Motivation (B) Habituation
(C) Imprinting (D) Fixed pattern learning

14. A state of reduced metabolism that occurs in animals living under intense heat is called :

(A) Hibernation (B) Dormancy (C) Estivation (D) Exothermy

15. A behaviour that mobilises your body for greater activity is called :

(A) Aggression (B) Reflex
(C) Fight and flight response (D) Instinct

16. Ivan Pavlov was first to demonstrate :

(A) Insight in dogs (B) Habituation in dogs
(C) Conditioning in dogs (D) Imprinting in dogs

17. A state of reduced metabolism that occur in extreme cold environment is called as :

(A) Habituation (B) Hibernation
(C) Dormancy (D) Estivation

18. In ...., some animals are subordinate to others :

(A) A dominance hierarchy (B) Competition
(C) Aggression (D) Agonistic

19. Territoriallity reduces :

(A) Instinctive behaviour (B) Competition
(C) Learning (D) Communication

20. Trial and error learning is also known as :

(A) Operant conditioning (B) Classical conditioning
(C) Insight learning (D) Innate behaviour

21. Any change in which a behaviour is altered as a result of experience is called :

(A) Habituation (B) Innate behaviour
(C) Learning (D) Imprinting

22. Glands and the hormones they produce are known to affect behaviours. In humans, which gland and hormone are responsible for controlling the active versus sleepy circadian rhythm :

(A) Pituitary gland and oxytocin (B) Pineal gland and melatonin
(C) Pancreas and insulin (D) Parathyroid gland and PTH

23. Which of the following behaviours would be classified as a reflex :

(A) Blinking of eyes as an insect flies towards your face
(B) A spider spinning a web (C) A dog barking at a stranger
(D) Tongue rolling by a cow

24. Male silverback gorillas are known to come to the defense of members of their group. In some cases, the large silverbacks will lose their own lives in the defense of a group member. This is an example of :

(A) Dominancy (B) Altruism
(C) Reciprocal altruism (D) Taxis

25. Why is the behaviour called a fixed action pattern considered to be "fixed"
(A) The behaviour is "fixed" in the animal's brain by receiving a positive reward each time the behaviour is done.
(B) The animal will perform the behaviour only at a certain time during its natural circadian rhythm.
(C) Once the animal begins the behaviour, it will complete it even if the stimulus that caused the behaviour is removed.
(D) The behaviour is learned or "fixed" on the organism by observing her parents repeat the behaviour many times while she was young.

26. The amount of gastric fill has been mediated by .... receptors in the stomach wall :
(A) Stretch receptor (B) Osmoreceptor
(C) Epithelial receptor (D) All of the above

27. Humoral satiety center involves all the following hormone except :
(A) Cholecystokinin (B) Ghrelin
(C) Leptin (D) Somatostatin

28. Which statement is false about leptin :
(A) Animal with homozygous recessive gene (ob/ob), becomes obese
(B) Leptin acts by inhibiting the orexic effect of neuropeptide Y (NPY)
(C) Leptin plays crucial role in reproduction
(D) Leptin acts by inhibiting the melanocortin system

29. Long term memory requires :
(A) Protein synthesis for formation of new communication between neurons
(B) Simply passage of impulses from one neuron to other
(C) Both (A) and (B)
(D) None of the above

30. Which statement is false about learning :
(A) Classical conditioning involves unconditioned stimulus which elicit response
(B) Operant conditioning also involves unconditioned stimulus
(C) Conditioned taste aversion occurs in response to nausea
(D) Spatial learning gives the ability to memorize the route and recognize the surrounding

31. Circadian rhythm is usually set by some external factors called as :
(A) Jet lag (B) Pacemaker (C) Zeitgeber (D) Ultradian

32. Sexual cycles are examples of :
(A) Circadian rhythm (B) Infradian rhythm
(C) Ultradian rhythm (D) Diurnal rhythm

33. When animal is being exposed to different to a different rhythm then .... results.
(A) Big lag (B) Time lag (C) Jet lag (D) Small lag

34. Acute distress is best measured by :
(A) Parasympathetic nervous system (B) Sympathetic nervous system
(C) Enteric nervous system (D) Somatic nervous system

35. Chronic stress is best measured by :

(A) Cortisol (B) Epinephrine
(C) Norepinephrine (D) Aldosterone

36. Which of the following is not used as T cell mitogen :

(A) Con.canavalin (B) Phytohemagglutin.A
(C) Poke weed (D) Peptidoglycan

37. Repetitive, apparently functionless behaviour is called :

(A) Repetitive behaviour (B) Stereotypic behaviour
(C) Abnormal behaviour (D) Motivational behaviour

38. What is false about stereotypic behaviour :

(A) Repetitive, apparently functionless behaviour
(B) Abnormality commonly indicate adaptation to poor environment
(C) Punish the animal (D) Change the environment

39. Behavioral technique used to evaluate the environment includes :

(A) Observation (B) Preference test
(C) Motivational test (D) All of the above

40. Dogs showing aggression towards people is a behavioural problem may be due to :

(A) Due to mutation or deletion of some genes
(B) The breakdown product serotonin, dopamine like neurotransmitter are lower in CSF
(C) Territorial aggression
(D) All of the above

41. Separation anxiety is commonly seen in :

(A) Cows (B) Pigs (C) Dogs (D) Rodents

42. Canine cognitive dysfunction can be treated by :

(A) Monoamine oxidase (MAO) (B) Homovanillic acid
(C) Inhibitors of MAO (D) Tryptophan

43. Which is false about glucostatic theory :

(A) Reduced blood glucose level limits glucose supply to brain resulting a signal for appetite
(B) Increase blood glucose increases brain glucose supply which in turn reduces food intake
(C) Proposed by Jean Mayer (D) Both (B) and (C)

44. Which theory is related to appetite control :

(A) "Lipostatic theory" established by Gordon C. Kennedy
(B) "Glucostatic theory" developed by Jean Mayer
(C) "Selfish Brain" theory by Achim Peters
(D) All of the above

45. Which statement is false about Flehman's response :

(A) was discovered by Frederik Ruysch and described later by Ludwig Jacobson
(B) main function of flehmen is intraspecific communication by transferring air containing pheremones

(C) it is a posture where animal curls back its upper lips exposing its front teeth with the nostrils usually closed

(D) Only males exhibit Flehman's response

46. Which statement is false about female sexual behaviour :

(A) Proceptive behaviour is attraction to male

(B) Receptive behaviour is attraction to male

(C) Estrogen stimulates proceptive and progesterone stimulates receptive behaviour

(D) During receptive behaviour female will stand for being mounted

47. Estrogen influence behaviour of estrus cow via :

(A) Sexual behaviour (B) Increase in activity

(C) Increase in feeding (D) All of the above

48. Estrogen increases the sensitivity of brain to the satiety hormone :

(A) CCK (B) Ghrelin (C) Leptin (D) Secretin

49. Androstenol is a pheromone present in................ of boar.

(A) Urine (B) Prepucial secretion

(C) Saliva (D) All of these

50. Maternal behaviour is investigated till date in more depth in which domestic species :

(A) Cow (B) Ewe (C) Sow (D) Doe

51. Maternal behaviour can be induced in multiparous nonpregnant ewes by following which regimen :

(A) Vaginal cervical stimulation (B) Increase estrogen level

(C) Decrease progesterone level (D) All of the above

52. Exclusivity is a maternal behaviour is due to :

(A) Oxytocin (B) Prolactin (C) Estrogen (D) Progesterone

53. Which statement is false about maternal behaviour :

(A) Mismothering is most likely occur when groups of female are confined

(B) Rejection, abandonment and aggression towards neonate is common in primiparous females

(C) Exclusivity of maternal behaviour is due to prolactin

(D) Segregation, nest building, vocalization, licking are different maternal behaviours

54. The process by which an ewe accept an alien lamb is called :

(A) Exclusivity (B) Pseudopregnancy

(C) Concaveation (D) Maternity

55. Which statement is wrong about grazing behaviour of cattle :

(A) Cattle tend to graze facing into the wind

(B) Cattle will not graze over their own faeces

(C) Cattle use their tongues to pull forage into their mouths before biting it off

(D) Cattle cannot graze in night

56. The "Wisdom of the body" is written by :

(A) W. B. Cannon (B) Ivan Pavlov

(C) Claude Bernard (D) Gordon C. Kennedy

57. Satiety center is present at :

(A) Anterior hypothalamus (B) Posterior Hypothalamus
(C) Ventro medial Hypothalamus (D) Lateral Hypothalamus

58. Lordosis is a sexual behaviour exhibited by :

(A) Male (B) Female
(C) Both (A) and (B) (D) Birds

59. when male pheromones stimulate synchronous estrus in a female population :

(A) Whitten effect (B) Bruce effect
(C) McClintock effect (D) Vandenbergh effect

60. An early induction of the first estrous cycle in prepubertal female mice exposure to the pheromone laden urine of a sexually mature (dominant) male mouse :

(A) Whitten effect (B) Bruce effect
(C) McClintock effect (D) Vandenbergh effect

61. Suppression or prolongation of oestrous cycles of mature female mice, when they are housed in groups and isolated from males :

(A) Bruce effect (B) McClintock effect
(C) Vandenbergh effect (D) Lee-Boot effect

62. Menstrual synchrony, also called :

(A) Bruce effect (B) McClintock effect
(C) Vandenbergh effect (D) Lee–Boot effect

63. Tendency of female rodents to terminate their pregnancies following exposure to the scent of an unfamiliar male at a critical period of pregnancy :

(A) Bruce effect (B) McClintock effect
(C) Vandenbergh effect (D) Lee-Boot effect

64. During estrus period rumination in cattle :

(A) Increases (B) Decreases
(C) Not affected (D) Ceases

65. Some slow breeders show marked improvement when exposed to breeding activity of other horses :

(A) Novelity effect (B) Voyeur effect
(C) Anxiolytic effect (D) All of the above

66. Among several social hierarchies which one has one or two triangular relationship :

(A) Linear (B) Linear tending
(C) Complex (D) Pecking order

67. In which upper clefted lip helps in close grazing :

(A) Sheep (B) Goat (C) Cattle (D) Horse

68. Endothermy :

(A) is only seen in mammals.
(B) involves production of heat through metabolism.
(C) is only seen in insects.
(D) is a term equivalent to "cold-blooded."

69. Selye's General adaptation syndrome (GAS) has 3 different phases except :
(A) Alarm phase (B) Fight and flight stage
(C) Resistance phase (D) Exhaustion stage

70. Thirst center is present at :
(A) Medulla (B) Hypothalamus
(C) Median eminence (D) Pons

71. In a social group when relative ranking is not assigned/ where all members are equal is called :
(A) Despotism (B) Egalitarian
(C) Pecking order (D) None of these

72. Minimum distance of approach to an animal before it flees is called :
(A) Alert zone (B) Flight distance
(C) Social distance (D) Individual space

73. Nest building is a well marked maternal behaviour in :
(A) Cow (B) Ewe
(C) Sow (D) Doe

74. Licking is a maternal behaviour helpful for all of the following except :
(A) Stimulates breathing, circulation, urination and defecation
(B) Dries up calf's coat which prevents evaporative heat loss
(C) Improves hygiene and reduces risk of predator attack
(D) Improves Flehman's response

75. Emotion is mainly control by :
(A) Cerebellum (B) Limbic system
(C) Pons (D) Medulla

76. Site of memory formation is :
(A) Hippocampus (B) Amygdala
(C) Hypothalamus (D) Pons

77. .... is a key organ involved in emotional processes.
(A) Hippocampus (B) Amygdala
(C) Hypothalamus (D) Medulla

78. Factors that affect circadian rhythms include :
(A) Only dark and light cycle (B) Nutrition and social cues
(C) Hormone feedback mechanism (D) All of the above

79. Cortisol levels were highest in blood of animal :
(A) In the morning (B) In the noon
(C) In the evening (D) In the mid night

80. Field of biology that study the biological rhythms in living organisms is called :
(A) Rhythmic biology (B) Sync biology
(C) Chronobiology (D) Jet biology

81. Domestic hens show unique sleeping style called :
(A) Monophasic sleep (B) Biphasic sleep
(C) Unihemispherical sleep (D) Bihemispherical sleep

82. Appetite center is present at :
(A) Anterior hypothalamus (B) Ventromedial nucleus
(C) Lateral hypothalamus (D) Posterior hypothalamus

83. Among the following .... is a long day breeder.
(A) Buffalo (B) Ewe
(C) Mare (D) Bitch

84. Melatonin is a stimulator for reproductive behaviour in :
(A) Long day breeder (B) Short day breeder
(C) In both (D) Not a stimulator at all

85. An abnormal behaviour mostly seen in high producing cows, where they behave like a bull :
(A) Homo sexuality (B) Nymphomania
(C) Buller steer syndrome (D) Masturbation

86. Major route of heat loss in birds is :
(A) Kidney (B) Skin
(C) Lungs (D) Both (A) and (B)

87. What is peck order :
(A) Expression of dominancy in birds (B) Ranking of animals
(C) Similar to Flight distance (D) All of the above

88. The distance at which animal will alert to the approaching threat :
(A) Alert zone (B) Flight distance
(C) Social distance (D) Individual space

89. Which among the following is a flocking behaviour :
(A) Grazing behaviour (B) Aggression
(C) Both (A) and (B) (D) None of these

90. In cattle shedding of hair occur in spring in a characteristic pattern. It starts on the :
(A) Head, neck and breech, midline of back and spread to each side of body
(B) Midline of back, spread to each side, head, neck and then breech
(C) Lower part of body, head neck and breech, midline of the back and then side of the body
(D) None of these pattern

91. In a temperate climate under normal conditions, the greatest loss of body heat occurs through :
(A) Conduction (B) Convection
(C) Radiation (D) Evaporation

92. .... is the center for integrating thermal information and regulating body temperature.
(A) Thalamus (B) Hypothalamus
(C) Median eminence (D) Pineal gland

93. .... stimulates sweating rates.
(A) sympathetic nervous system (B) parasympathetic nervous system
(C) enteric nervous system (D) Both (A) and (B)

94. Eccrine glands secrets :

(A) Secretes a viscous milky fluid (B) Secretes watery fluid

(C) Secrets enzyme rich fluid (D) Secrets bloody fluid

95. Which statement is correct :

(A) Heat is lost more readily from the body surface during water immersion than in air

(B) The thalamus is the center for integrating thermal information

(C) Vasoconstriction of cutaneous vessels promotes heat loss from the body

(D) All of the above are true

96. Which of the following statements regarding basal metabolic rate is correct :

(A) Basal metabolic rate is defined as the lowest metabolic rate that occurs in any 24 hour period

(B) The basal metabolic rate is increased in people suffering from hypothyroidism

(C) Basal metabolic rate increases with age

(D) Adult males have a higher basal metabolic rate than adult females

97. Which of the following statement is correct :

(A) The respiratory quotient is the amount of oxygen used divided by the amount of carbon dioxide produced.

(B) The oxygen consumption of a normal adult is a good measure of the metabolic rate

(C) Metabolic rate increases slightly when walking slowly compared to sitting

(D) During starvation, metabolic rate increases as protein is broken down to make glucose

98. Homeostasis :

(A) refers to the unwavering control of a physiological setpoint

(B) refers to maintaining a stable internal environment

(C) refers to maintaining a stable external environment

(D) Both (A) and (B)

99. The term "metabolism" :

(A) refers to all the chemical reactions that occur in the body

(B) includes only the synthesis of complex molecules from simpler molecules

(C) includes only the breakdown of complex molecules into simpler molecules

(D) none of the above

100. Glycolysis :

(A) yields two moles of ATP for each mole of glucose processed

(B) may yield two moles of lactate for each mole of glucose processed

(C) takes place in the mitochondria of cells

(D) Both (A) and (B)

101. The resting membrane potential :

(A) occurs only in nerve and muscle cells

(B) is the same in all cells

(C) is oriented so that the cell's interior is positive with respect to the extracellular fluid

(D) occurs in all cells

102. Those who can regulate their own internal body temperature, regardless of fluctuating external temperatures are :

(A) Endotherms (B) Ectotherms
(C) Poikilotherns (D) Heterothems

103. Evaporative heat loss is minimum :

(A) Within thermoneutral zone (B) Beyond thermoneutral zone
(C) Both (A) and (B) (D) Within zone of homeothermy

104. Which of the following is the energy efficient method of evaporative heat loss :

(A) Panting (B) Sweating
(C) Saliva spreading (D) Both(A) and (B)

105. Which among the following is a good sweated :

(A) Cattle (B) Horse (C) Dog (D) Buffalo

106. During ovulation body temperature :

(A) Decreases 0.5 to 10 degree (B) Increases 0.5 to 10 degree
(C) Does not alter (D) Decreases 1.5 to 20 degree

107. Growth curve is sigmoid shape except in :

(A) Man (B) Horse (C) Cattle (D) Pig

108. Growth inhibitors and growth retarders are called as .... respectively.

(A) Promine, Inhibin (B) Promine, Retine
(C) Inhibin, Promine (D) Retine, Promine

109. Surface law is given by

(A) Broady (B) Rubner
(C) Kata (D) Walter B. Cannon

110. Psychrometer is used to measure :

(A) Minimum and maximum environmental temperature
(B) Atmospheric pressure
(C) Relative humidity
(D) Temperature and humidity simultaneously

111. Which type of respiration leads to alkalosis :

(A) Decrease respiration
(B) Increase Deep bellow respiration
(C) Increase Shallow panting respiration
(D) Both (A) and (B)

112. Which among the following expression accurately explain relationship between surface area and weight of animal :

(A) Surface area = $0.15\ W^{0.56}$ (B) Surface area = $0.25\ W^{0.76}$
(C) Surface area = $0.35\ W^{0.56}$ (D) Surface area = $0.45\ W^{0.76}$

113. Which among the following does not indicate true growth :

(A) Hyperplasia (B) Hypertrophy
(C) Accumulation of adipose tissue (D) Accumulation of cellular DNA

114. Maximum renewing cell population is present in which tissue :

(A) Epidermis (B) Mesoderm
(C) Skeletal tissue (D) Connective tissue

115. Body weight in young ruminants can be calculated by :

(A) Shaffers formula (B) Aggarwals formula
(C) Both (A) and (B) (D) None of these

116. Warm bloodedness is a manifestation of high rate of heat production which may be about .... times higher in homeotherms than in poikilotherms of same body size and at the same body temperature.

(A) 2-3 (B) 4-5 (C) 6-7 (D) 8-10

117. Summer stress in cattle can be reduced by

(A) Sprinkling water on the body
(B) Tethering at shady area
(C) Providing cool drinking water ad libitum
(D) All of the above

— Answer Sheet —

Q. No.	Ans	Q. No.	Ans	Q. No.	Ans	Q. No.	Ans	Q. No.	Ans	Q. No.	Ans
1.	C	21.	C	41.	C	61.	D	81.	C	101.	D
2.	D	22.	B	42.	C	62.	B	82.	C	102.	A
3.	B	23.	A	43.	B	63.	A	83.	C	103.	A
4.	B	24.	B	44.	D	64.	B	84.	B	104.	B
5.	D	25.	C	45.	D	65.	B	85.	B	105.	B
6.	C	26.	A	46.	B	66.	B	86.	C	106.	B
7.	C	27.	D	47.	D	67.	D	87.	A	107.	A
8.	A	28.	D	48.	A	68.	B	88.	A	108.	D
9.	A	29.	A	49.	D	69.	B	89.	A	109.	B
10.	C	30.	B	50.	B	70.	B	90.	A	110.	C
11.	D	31.	C	51.	D	71.	B	91.	C	111.	B
12.	D	32.	B	52.	A	72.	B	92.	B	112.	A
13.	C	33.	C	53.	C	73.	C	93.	A	113.	C
14.	C	34.	B	54.	C	74.	D	94.	B	114.	A
15.	C	35.	A	55.	D	75.	B	95.	A	115.	D
16.	B	36.	D	56.	A	76.	A	96.	D	116.	B
17.	B	37.	B	57.	C	77.	B	97.	B	117.	D
18.	A	38.	C	58.	B	78.	D	98.	B		
19.	B	39.	D	59.	A	79.	A	99.	A		
20.	A	40.	D	60.	D	80.	C	100.	D		

• Chapter 2 •

Blood and Circulatory System

1. **'Clear watery fluid which comes out after clot retraction is called :**
   (A) Serum (B) Plasma
   (C) Pus (D) Heparin
2. **What percent of blood volume is plasma?**
   (A) 15 (B) 35 (C) 45 (D) 65
3. **Plasma contains......... percent water.**
   (A) 91-92 (B) 65-75 (C) 45-55 (D) 20-30
4. **Which is true about plasma protein :**
   (A) Globulin is the key plasma protein for regulation of colloidal osmotic pressure
   (B) Globulin includes immunoglobulin that functions in immune response
   (C) Fibrinogen is the precausor of fibrin
   (D) Albumin and globulin also important transporter of many substance.
5. **RBC in camel :**
   (A) Elliptical and anucleated (B) Oval and anucleated
   (C) Elliptical and nucleated (D) Oval and nucleated
6. **RBC in bird :**
   (A) Elliptical and anucleated (B) Oval and anucleated
   (C) Elliptical and nucleated (D) Oval and nucleated
7. **Smallest RBC is seen in :**
   (A) Dog (B) Cat (C) Sheep (D) Goat
8. **Largest RBC is seen in**
   (A) Dog (B) Cat (C) Sheep (D) Goat
9. **Erythropoiesis in adult life occurs at**
   (A) Lymph node (B) Spleen
   (C) Bone marrow (D) Liver
10. **RBC contains .... percent solid.**
    (A) 7-8 (B) 15 (C) 35 (D) 75
11. **.................. is the principal cation of erythrocyte.**
    (A) Sodium (B) Potassium
    (C) Magnesium (D) Calcium
12. **Red colour of RBC is due to :**
    (A) Bilirubin (B) Haemoglobin
    (C) Carotene (D) Xanthophilin
13. **Blood volume accounts .... percent of lean body weight :**
    (A) 5 (B) 8-10 (C) 10-20 (D) 15-30
14. **Abnormal size of RBC is called :**
    (A) Anisocytosis (B) Poikilocytosis
    (C) Aplasocytosis (D) Spherocytosis

15. What is the advantage of increased biconcavity of RBC :
(A) Decrease diffusion distance (B) Increase surface area
(C) Decrease osmotic frazility (D) All of these

16. Extramedullary hematopoiesis occurs at :
(A) Marrow of long bone (B) Membranous bone
(C) Spleen and liver (D) Both (B) and (C)

17. Hemoglobin synthesis begins at .... stage of erythropoiesis.
(A) Reticulocyte (B) Rubricyte
(C) Rubriblast (D) Metarubricyte

18. Which one does not belongs to myeloid series :
(A) Neutrophil (B) Eosinophil
(C) Monocyte (D) Lymphocyte

19. Cell membrane of RBC contains .... protein which carries blood group antigen.
(A) Glycophorin (B) Spectrin
(C) Glucophorin (D) Rh protein

20. Cell membrane of RBC contains .... protein which is contractile in nature.
(A) Glycophorin (B) Spectrin (C) Actin (D) Rh protein

21. Time taken for rubriblast to become RBC is
(A) 3 days (B) One week (C) Two week (D) 24 hours

22. RBC count in cattle is :
(A) 2 million/ ul (B) 3 million/ ul
(C) 5 million/ ul (D) 7 million/ ul

23. Intravascular hemolysis accounts .... percent of total lysis of RBC.
(A) 10 (B) 20 (C) 30 (D) 40

24. Site of RBC destruction in domestic animal is :
(A) Liver (B) Spleen
(C) Bone marrow (D) Kidney

25. Site of RBC destruction in birds is :
(A) Liver (B) Spleen (C) Bone marrow (D) Kidney

26. Lifespan of RBC in domestic animals :
(A) 20-30 days (B) 50-60 days
(C) 100-120 days (D) 300 days

27. Lifespan of RBC in birds is :
(A) 20-30 days (B) 50-60 days
(C) 100-120 days (D) 300 days

28. Mean corpuscular haemoglobin concentration decreases in :
(A) Iron deficient anemia
(B) Vitamin B 12 deficient anemia
(C) Pyridoxine deficient anemia
(D) Riboflavin deficient anemia

29. Heme synthesis starts and ends in :
(A) Mitochondria (B) Cytosol
(C) Golgi apparatus (D) Ribosome

30. Which is false about haemoglobin :
(A) Iron in haemoglobin is in ferric form
(B) Heme synthesis occurs at mitrochondria
(C) Hemoglobin synthesis starts at rubricyte stage
(D) Hemoglobin has 4 iron atoms

31. Fetal haemoglobin has more oxygen affinity compared to adult haemoglobin except in :
(A) Horse (B) Dog (C) Cat (D) Goat

32. Hemoglobin percent in chicken is .... gm/dl.
(A) 15-20 (B) 12-15 (C) 10-12 (D) 6.5-9

33. Erythropoietin synthesis occurs at .... in dogs.
(A) Liver (B) Kidney
(C) Bone marrow (D) spleen

34. Which one is false about erythropoietin :
(A) At higher altitude erythropoietin secretion increases
(B) Erythropoietin stimulates committed stem cell mitotic activity
(C) Half life of erythropoietin is one day
(D) Erythropoietin synthesis mainly occurs at liver

35. Largest cell of blood :
(A) Neutrophil (B) Monocyte
(C) Lymphocyte (D) Platelets

36. Which of the following statements about erythrocytes is correct :
(A) They fight infection (B) They clot blood
(C) They lack a nucleus (D) They are produced in the spleen

37. The formation of a blood clot is known as which of the following :
(A) Coagulation (B) Leucopoiesis
(C) Erythropoiesis (D) Chemotaxis

38. Platelets are formed from what type of cell :
(A) Myelocyte (B) Melanocyte
(C) Myocyte (D) Megakaryocyte

39. Which of the following decrease platelet aggregation & cause vasodilatation :
(A) Thromboxane A2 (B) Prostaglandin F2 alpha
(C) Prostaglandin E (D) Prostaglandin I

40. Secretory immunoglobulin is :
(A) Ig G (B) Ig A (C) Ig M (D) Ig E

41. The ratio of RBC to WBC in cattle is
(A) 1:8 (B) 1:80 (C) 1:800 (D) 1:8000

42. Blood circulation was first explained by :
(A) William Harvey (B) Mendel
(C) Jenner (D) Pasture

43. The protein responsible for iron transport in plasma is :

(A) Hemoglobin (B) Apo-transferrin

(C) Ceruloplasmin (D) Alpha2 antitrypsin

44. Prothrombin level falls in the blood due to lack of :

(A) Vitamin B12 (B) Vitamin K

(C) Platelate (D) Phospholipid

45. Vitamin B12 deficiency leads to :

(A) Microcytic hypocromic anemia (B) Hemolytic anemia

(C) Pernicious anemia (D) Megaloblastic anemia

46. A blood smear shows hypochromic, microcytic RBCs with moderate Poikilocytosis. Most likely diagnosis is :

(A) Anemia due to chronic diseases (B) Autoimmune hemolytic anemia

(C) Folate deficiency (D) Iron deficient anemia

47. A graphic curve representing the variation in size of red blood cells is called :

(A) Isodose curve (B) Curve of occlusion

(C) Curve of Carus (D) Price Jones curve

48. The plasma proteins are not involved in

(A) Blood clotting (B) Blood viscosity

(C) Colloidal osmotic pressure (D) Transport of oxygen

49. During intrauterine life, erythropoiesis occurs at :

(A) Liver (B) Spleen

(C) Yolk sac (D) Bone marrow

50. One molecule of haemoglobin contains :

(A) One heme + one polypeptide chain

(B) Two heme + two polypeptide chain

(C) One heme + four polypeptide chain

(D) Four heme + four polypeptide chain

51. Inside the body blood does not coagulate due to presence of :

(A) Fibrin (B) Heparin

(C) Hemoglobin (D) Thromboplastin

52. Blood pressure is measured by :

(A) Electrocardiogram (B) Stethoscope

(C) Hygrometer (D) Sphygmomanometer

53. Erythrocyte contains .... percent water.

(A) 60-70 (B) 70-80 (C) 80-90 (D) 90-95

54. Normal RBC number in bovines is

(A) 18 million (B) 15 million (C) 14 million (D) 7 million

55. Erythrocyte surface area in mammals varies from 57.67 m2 /Kg, whereas in birds it is :

(A) 100 m^2 /Kg (B) 88 m^2 /Kg (C) 77 m^2 /Kg (D) 44 m^2 /Kg

56. Vitamin B12 helps in RBC maturation it contains one molecule of :

(A) Magnesium (B) Copper (C) Cobalt (D) Iron

57. Iron in heme is in :

(A) Ferric form (B) Ferrous form
(C) Both (A) and (B) (D) Either one

58. Azurophilic granules of neutrophils are also called :

(A) Primary granule (B) Secondary granule
(C) Tertiary granule (D) Specific granule

59. Higher concentration of 2-3 bisphosphoglycerate (BPG) shifts the oxygen haemoglobin dissociation curve :

(A) Left (B) Right (C) Steeper (D) Unaltered

60. One gram haemoglobin can carry .... ml of oxygen in fully saturated stage.

(A) 1ml (B) 1.34ml (C) 1.44ml (D) 1.54ml

61. Muscle haemoglobin is also called :

(A) Myoglobin (B) Hemoglobin
(C) Musculoglobin (D) Metaglobin

62. The oxygen haemoglobin dissociation curve in muscle haemoglobin is :

(A) S shaped (B) Circular (C) Hyperbolic (D) Oval

63. Affinity of haemoglobin for carbon monoxide is .... times higher than oxygen.

(A) 10 times (B) 20 times (C) 100 times (D) 200 times

64. When carbon monoxide combines with haemoglobin, it is called :

(A) Carbaminohaemoglobin (B) Carboxyhaemoglobin
(C) Carbonmonoxyhaemoglobin (D) Both (B) and (C)

65. The true oxide of haemoglobin is :

(A) Carboxyhaemoglobin (B) Oxyhaemoglobin
(C) Methaemoglobin (D) Myoglobin

66. Methaemoglobin is reduced to haemoglobin by intravenous injection of :

(A) Methylene blue (B) Nitrite (C) Chlorate (D) Sulfonamides

67. In allergic condition and parasitic infection which leukocyte number increases significantly :

(A) Neutrophil (B) Eosinophil (C) Basophil (D) Lymphocyte

68. Which leukocyte has receptor for immunoglobulin (Ig E) :

(A) Neutrophil (B) Eosinophil (C) Basophil (D) Lymphocyte

69. What is false about interferons :

(A) A group of protein with antiviral activity
(B) Can be produced by lymphocytes
(C) They cannot inhibit tumor growth
(D) Interferon α and β are produced by B lymphocyte

70. Shift to left is used in differential leukocyte count (DLC) for :

(A) Increase in segmented neutrophils
(B) Increase in band neutrophils
(C) Decrease in both segmented neutrophils
(D) Decrease in band neutrophils

71. **Smallest blood cell :**
(A) Erythrocyte (B) Leukocyte
(C) Thrombocyte (D) Megakaryocyte

72. **The platelet count in normal domestic animals is :**
(A) 1-3 lakh (B) 3-6 lakh (C) 6-9 lakh (D) 9-12 lakh

73. **Normal lifespan of platelets is :**
(A) 2-3 days (B) 5 days (C) 8-10days (D) 15 days

74. **Largest lymphoid organ of animal body is**
(A) Spleen (B) Lymph node (C) Thymus (D) Kidney

75. **B lymphocyte accounts .... percent of total circulating lymphocyte in blood stream.**
(A) 15 (B) 35 (C) 55 (D) 85

76. **Specific gravity of whole blood is about :**
(A) 1.014 (B) 1.024 (C) 1.053 (D) 1.089

77. **ESR (erythrocyte sedimentation rate) is generally measured by :**
(A) ml/hr. (B) mm/hr. (C) cm/hr. (D) mm/min.

78. **ESR is higher in :**
(A) Cattle (B) Sheep (C) Dog (D) Horse

79. **Most abundant immunoglobulin in most domestic animals is :**
(A) Ig G (B) Ig M (C) Ig A (D) Ig E

80. **The primary immunoglobulin in cow colostrum and milk is :**
(A) Ig G (B) Ig M (C) Ig A (D) Ig E

81. **Plasma albumin originate from :**
(A) Mucosal tissue (B) Liver
(C) Kidney (D) Diet

82. **Colloidal osmotic pressure due to plasma proteins (oncotic pressue) is about :**
(A) 3 mm Hg (B) 8 mm Hg (C) 18 mm Hg (D) 28 mm Hg

83. **Albumin normally accounts .... percent of oncotic pressure.**
(A) 60 (B) 70 (C) 80 (D) 90

84. **Which one among the following is the predominant cell in circulating blood after prolonged period of erythropoietin secretion :**
(A) Eosinophil (B) Neutrophil
(C) Reticulocyte (D) Platelets

85. **What is the normal WBC count in bovines :**
(A) 2000 WBC/μl (B) 4000 WBC/ μl
(C) 7000 WBC/ μl (D) 15000 WBC/ μl

86. **Acidosis will shift oxygen.haemoglobin dissociation curve :**
(A) to the right and makes the oxygen more readily available to the tissues
(B) to the right and makes the oxygen less readily available to the tissues
(C) to the left and makes the oxygen more readily available to the tissues
(D) to the left and makes the oxygen less readily available to the tissues

87. which one is false regarding extra.medullary hematopoiesis :

(A) It occurs in severe hemolytic anemia
(B) It occurs in megaloblastic anemia
(C) It occurs in secondary polycythemia
(D) It occurs in fetus

88. Which one of the following is true about sickle cell trait :

(A) It is caused due to anemia
(B) It protects against malaria
(C) It occurs mainly in females
(D) It is a cause of frequent sickle cells in peripheral blood

89. Hemophilia A is an inherited coagulation defect due to deficiency of :

(A) Factor II (B) Factor IV (C) Factor VIII (D) Factor IX

90. The normal haemostatic response to vascular damage depends on three major factors. Which ONE of these is not included in that group?

(A) The blood vessel wall (B) Stasis of the blood flow
(C) Circulating platelets (D) Blood coagulation factors

91. Which ONE of these statements is TRUE concerning von Willebrand factor :

(A) It cross - links platelets to each other
(B) It carries factor IX
(C) After release it forms large aggregates that are needed for its function
(D) Plasma vWF is derived from endothelial cells

92. Which ONE of these statements concerning the initiation of blood coagulation is NOT TRUE?

(A) Tissue factor is expressed on endothelial cells
(B) Coagulation is initiated by the interaction of tissue factor with plasma factor VIIa
(C) The factor VIIa.tissue factor complex activates both factor IX and factor X
(D) One to two per cent of the total factor VII circulates in the activated form, but does not express proteolytic activity unless bound to tissue factor

93. A hematocrit of 41% means that in the sample of blood analyzed :

(A) 41% of the haemoglobin is in the plasma
(B) 41% of the total blood volume is made up of blood plasma
(C) 41% of the total blood volume is made up of red and white blood cells and platelets
(D) 41% of the formed elements in blood are red blood cells

94. Factor X of blood clotting is called :

(A) Antihemophillic factor (B) Stuart's factor
(C) Hageman's factor (D) Christmas factor

95. Thromboxane A2 (TXA2) produced from :

(A) Neutrophil (B) RBC (C) Macrophage (D) Platelets

96. The key agent that converts prothrombin to thrombin is :

(A) Factor XII (B) Factor IX (C) Factor X (D) Factor III

97. Calcium is considered as .... coagulation factor.

(A) Factor I (B) Factor II (C) Factor III (D) Factor IV

98. Among the following which one is not an inhibitor of coagulation :

(A) Anti.thrombin
(B) Thrombomodulin – thrombin complex
(C) Protein C and protein S
(D) Thromboxane A2 (TXA2)

99. Fibrinogen is converted to fibrin by removal of :

(A) Fibrinopeptide A (B) Fibrinopeptide B
(C) Fibrinopeptide C (D) Both (A) and (B)

100. In sickle cell anemia :

(A) Insufficient production of α chain
(B) Insufficient production of β chain
(C) Amino acid valine is replaced for glutamic acid
(D) Amino acid glutamic acid is replaced for valine

101. Which one is false about blood doping :

(A) It is the process of increasing RBC number
(B) Possible by transfer of homologous RBCs
(C) First time used illegally in race horse
(D) Cannot be done by use of synthetic erythropoietin

102. Which one is false about heterophils :

(A) Present in avian species
(B) It stains neutral dye
(C) It is a leukocyte similar to neutrophil
(D) Ratio of heterophils to lymphocyte is a good indicator of stress

103. Hill's coefficient of myoglobin is 1 but of haemoglobin is mammals is :

(A) 1 (B) 2.4 (C) 3.5 (D) 4

104. Which one is false about hemal node/ hemolymph node :

(A) Found only in ruminants
(B) Found only in avian species
(C) Similar in structure function with spleen
(D) Erythropoiesis occurs here in fetal period

105. Transferrin is an example of :

(A) α 1 globulin (B) α 2 globulin
(C) β globulin (D) γ globulin

106. Neutrophil extracellular trap (NET) is a unique method of antimicrobial action discovered by :

(A) Brinkman and Zyclinsky (B) Aschoff and Landau
(C) Haldane and Pristley (D) Hamberger

107. What potentiate the action of heparin :

(A) Thrombin (B) Fibrin
(C) α 2 anti-trypsin2 (D) anti-thrombin III

108. Normal ESR of horse is :

(A) 2-4 mm/hr (B) 2-12 mm/min
(C) 2-4 mm/7 hr (D) 0-3 mm/hr

109. Ternary enzyme complex formation occurs during fibrin degradation, inclues all except :

(A) Plasminogen (B) Fibrin

(C) Thrombomodulin (D) Tissue plasminogen activator (t PA)

110. Tense complex of intrinsic pathway in coagulation composed of all except :

(A) Factor IV (B) Factor VII (C) Factor VIII (D) Factor IX

111. Prothrombinase complex needed for conversion of prothrombin to thrombin includes all except :

(A) Factor IV (B) Factor V (C) Factor IX (D) Factor X

112. Pulmonary circulation is also called :

(A) Lesser circulation (B) Greater circulation

(C) Systemic circulation (D) All of the above

113. Which one is not included in microcirculation :

(A) Capillary (B) Venules

(C) Pre.capillary sphincter region (D) All of the above

114. Vascular smooth muscle is present all over the vascular tree except in :

(A) Arterioles (B) Veins (C) Capillaries (D) All of the above

115. Which one among the following is not a endothelium derived relaxing factor :

(A) Nitric oxide (B) Prostacyclin

(C) Endothelin (D) Both (A) and (B)

116. Windkessel vessel include :

(A) Aorta (B) Big arteries (C) Arterioles (D) Both (A) and (B)

117. Capacitance vessels are :

(A) Arteries (B) Veins (C) Capillaries (D) Aorta

118. Resistance vessels are

(A) Aorta (B) Arterioles (C) Veins (D) Capillaries

119. Fenestrated capillaries are present in :

(A) Blood brain barrier (B) Liver

(C) Kidney (D) Bone marrow

120. Mucosal block theory was given by .... in 1946.

(A) Granick (B) Fredrick (C) Harvey (D) Thomas Mayer

121. Exchange vessels are also called :

(A) Capillary (B) Lymphatics (C) Arterioles (D) Venules

122. Continuous capillaries are seen in :

(A) Brain (B) Liver (C) Kidney (D) Bone marrow

123. Minute amount interstitial fluid that is usually entrapped in the interstitium is also called :

(A) Filtered fluid (B) Tissue gel

(C) Cell gel (D) None of the above

124. Normally about .... fraction of filtered fluid cannot reabsorbed by the blood and returned via lymphatics.

(A) 1/10 (B) 1/30 (C) 1/50 (D) 1/80

125. Daily lymph formation varies from :
(A) 0.5-1 lit/day (B) 2-3 lit/day (C) 5-6 lit/day (D) 6-7lit/day

126. Lymph from lower part and left side of body is received by :
(A) Right lymphatic duct (B) Left lymphatic duct
(C) Thoracic duct (D) Both (B) and (C)

127. Average protein content of lymph is about :
(A) 0.5 g/dl (B) 1 g/dl (C) 3 g/dl (D) 8 g/dl

128. Which region of lymph in our body contains maximum protein :
(A) Kidney (B) Intestine (C) Spleen (D) Liver

129. Left ventricular failure leads to :
(A) Generalized edema (B) Pulmonary edema
(C) Intracellular edema (D) Liver cirrhosis

130. Left ventricular failure leads to :
(A) Generalized edema (B) Pulmonary edema
(C) Intracellular edema (D) Liver cirrhosis

131. Capillaries carry about .... percent of total blood in blood vessels.
(A) 5 (B) 15 (C) 25 (D) 80

132. In venous blood haemoglobin is .... percent saturated :
(A) 30 (B) 45 (C) 72 (D) 97.5

133. What among the following shifts the O2 – Hemoglobin dissociation curve to left
(A) High PH (B) High temperature
(C) High 2,3 BPG (D) None of the above

134. Fractional extraction of avian blood (for oxygen at tissue level) is ....
(A) 1/4 (B) 1/3 (C) 1/2 (D) 1

135. Camel can be swell upto .... percent of normal.
(A) 120 (B) 140 (C) 220 (D) 240

136. Reynolds number (Re) used to predict ....
(A) Resistance (B) Conductivity (C) Turbulence (D) Current

137. Flat disc shaped RBCs are present in :
(A) Horse and Dog (B) Goat and Dog
(C) Goat and Horse (D) Horse and Camel

138. Which among the following is a cellular growth regulator :
(A) Trypsin (B) Secretin (C) Erythropoitin (D) Bacitracin

139. Which is the first recognizable cell type by Leishman's stain among the following developmental stages of neutrophil :
(A) Myeloblasts (B) Promyelocyte
(C) Myelocyte (D) Band form

140. Basophils have similar features like .... found in tissue.
(A) Natural killer cells (B) Mast cells
(C) Phagocytes (D) Parietal cell

141. Basophils contain .... in their granules.
(A) Heparin (B) Histamine (C) Both (A) & (B) (D) RNA

142. Purpura is a condition where purple or red discolored spot appears on the skin is due to dietary deficiency of :
(A) Vitamin A (B) Vitamin B (C) Vitamin C (D) Iron and Zinc

143. Which among the following is true about Vit. B12 synthesis :
(A) Synthesized in human body
(B) Not synthesized in human body
(C) Bacteria cannot synthesize vitamin B12
(D) Both (A) and (C)

144. Which among the following is called myeloid growth factor as it stimulates WBC production :
(A) Interleukins (B) Stem cell factor
(C) Colony stimulating factor (D) All of the above

145. Increased Rouleaux formation or RBC piling resulting increase ESR occurs in:
(A) Increase of plasma fibrinogen (B) Increase of gamma globulin
(C) Infection and inflammations (D) All of the above

146. Which system in circulatory complex has high capacitance :
(A) Arterial system (B) Venous system
(C) Capillary system (D) Both (B) and (C)

147. Two capillary systems connected in series is called :
(A) Fenestrated capillary system (B) Perfusion system
(C) Portal system (D) All of the above

148. .... receives 100% of cardiac output.
(A) Liver (B) Lungs (C) Kidney (D) Skin

149. What is pulse pressure :
(A) Difference between systolic and diastolic pressure
(B) Difference between diastolic and systolic pressure
(C) One-third systolic pressure + Diastolic Pressure
(D) One-third Diastolic Pressure + systolic pressure

150. Stroke volume is :
(A) Amount of blood pumped out in each min
(B) Amount of blood pumped out in each sec
(C) Amount of blood pumped out in each heart beat
(D) Both (B) and (C)

151. The systemic arterial blood pressure is highest in :
(A) Horse (B) Giraffe (C) Bovine (D) Feline

152. Volume of blood comes out from both the ventricles per min. is called :
(A) Blood flow (B) Stoke volume
(C) Cardiac output (D) Blood volume

153. Which statement is false about venous blood vessels :
(A) Venous walls have less collagen and elastin compared to arterial wall
(B) Venous vessels can expand easily
(C) Venous vessels can hold large volume of blood
(D) All venous vessels contain deoxygenated blood

154. Which is true about pericardium :
(A) Pericardium is inelastic and it prevents acute expansion of heart
(B) Acute accumulation of pericardial fluid leads to collapse of venacava leading to poor cardiac filling resulting cardiac tamponade
(C) Pericardium has two layers
(D) All are true

155. The SA node is the normal pacemaker because of its :
(A) rate of impulse discharge (B) location in the atrium
(C) neural control (D) muscular structure

156. Which one is true about working myocardial cells :
(A) They have centrally located nucleus
(B) They have numerous mitochondria
(C) Sarcoplasmic reticulum has direct contact with cell exterior
(D) The cells are connected to each other at gap junction which is present on intercalated disc

157. Under normal circumstances major fuel for cardiac metabolism is :
(A) Carbohydrate (B) Free fatty acid
(C) Protein (D) Both (A) and (B)

158. In a resting cardiac muscle the membrane is in :
(A) Polarized condition (B) Depolarized condition
(C) Hyperpolarized condition (D) None of the above

159. Turbulence is enough to produce sound that is audible when the Reynolds number exceeds ....
(A) 200 (B) 1000 (C) 2000 (D) 3000

160. Which one among the following is not a function of atria :
(A) Acts as a pump, helps in ventricular filling
(B) Acts as elastic reservoir
(C) Assist in AV valve closure
(D) Assist in AV valve opening

161. Branch of physiology dealing with forces and physical mechanism concern with flow of blood is called :
(A) Rheology (B) Hemodynamics
(C) Circulatory physiology (D) Fluid dynamics

162. ....% of cardiac output is by atrial contraction.
(A) 20% (B) 30% (C) 50% (D) 80%

163. Relationship between left and right ventricular wall :
(A) Left ventricle is 3 times more thicker than right ventricle
(B) Right ventricle is 3 times more thicker than left ventricle
(C) Both right and left ventricles are of equal thickness
(D) The thickness varies with age

164. Pulse pressure is equal to :
(A) Diastolic pressur (B) Systolic pressue
(C) Diastolic pressure – systolic pressure
(D) Diastolic pressure + systolic pressure

165. Bulk flow occurs in :
(A) Aorta, Arteries and Arterioles
(B) Aorta, Arteries, Arterioles and capillaries
(C) Capillaries only (D) None of these

166. Cerebral infarction creates a condition called :
(A) Stoke (B) Cardiac tympanade
(C) Myositis (D) None of these

167. Arrangement of two systemic capillary bed is called :
(A) Multiple capillary (B) Multimodal capillary
(C) Portal capillary (D) Specialized capillary

168. Which one is true about cardiac output :
(A) The volume of blood pumped by each ventricle per minute is equal
(B) The volume of blood pumped per minute is unequal
(C) The volume of blood pumped by left ventricle is more
(D) Both (B) and (C)

169. Gates in circulatory system refers to :
(A) Aorta (B) Arteries (C) Veins (D) Arterioles

170. The fraction of cells in blood is called :
(A) Hematocrit (B) PCV (C) ESR (D) Both (A) and (B)

171. The two most common test for testing coagulation system are
(A) Prothrombin time and partial tromboplastin time :
(B) Fibrinogen time and prothrombin time
(C) ESR and prothrombin time
(D) Fibrinogen time and ESR

172. Continuous capillaries are found in :
(A) Brain (B) Liver (C) Kidney (D) Both (A) and (B)

173. Discontinuous capillaries are found in :
(A) Brain (B) Liver (C) Kidney (D) Both (A) and (B)

174. Fenestrated capillaries are found in :
(A) Brain (B) Liver (C) Kidney (D) Both (A) and (B)

175. In capillaries for passage of ions which is present :
(A) Intracellular cleft (B) Vesicular channel
(C) Carrier proteins (D) Both (A) and (B)

176. Vesicular channels are more in :
(A) Continuous capillaries (B) Discontinuous capillaries
(C) Fenestrated capillaries (D) Both (B) and (C)

177. Intracellular clefts are more in :
(A) Continuous capillaries (B) Discontinuous capillaries
(C) Fenestrated capillaries (D) Both (B) and (C)

178. Vasomotion of pre capillary sphincter is mainly regulated by :
(A) Oxygen concentration in tissue (B) Sodium concentration in tissue
(C) Positively charged ion concentration in tissue
(D) Both (A) and (B)

179. Fluid entrapped in minute spaces among the proteoglycan filaments of interstitium is called :

(A) Fluid gel (B) Tissue gel (C) Free fluid (D) Free gel

180. What is the normal proportion of filtered fluid that cannot reabsorbed back to blood directly and need accessory route to enter in to blood :

(A) 1/5th (B) 1/10th (C) 1/15th (D) 1/20th

181. What amount of lymph formation occurs daily in our body :

(A) 1 L/day (B) 2-3 L/day (C) 5-7 L/day (D) 7-9 L/day

182. Largest lymph duct of our body is called :

(A) Thoracic duct (B) Right lymph duct
(C) Left lymph duct (D) Both (B) and (C)

183. The average protein content of lymph is :

(A) 1g/dl (B) 3g/dl (C) 6g/dl (D) 8g/dl

184. The factors that affect lymphatic pump activity are :

(A) Contraction of surrounding skeletal muscle
(B) Movement of body parts
(C) Pulsation of artery adjacent to it
(D) All of the above

185. Heart failure leads to :

(A) Generalized edema (B) Pulmonary edema
(C) Both (A) and (B) (D) Liver cirrhosis

186. Lesser circulation is also called :

(A) Pulmonary circulation (B) Systemic circulation
(C) Cardiac circulation (D) Hepatic circulation

187. Which among the following is false about SA node :

(A) Present in right atrium
(B) Near the opening of superior venacava
(C) Called as accessory pace maker
(D) Conduct impulse to inter-aterial septum

188. Conduction cells of heart does not include :

(A) Working myocardial cells (B) Nodal tissue
(C) Bundle of His (D) Purkinje's Fibers

189. During cardiac diseases the repolarization phase of cardiac Action Potential is either interrupted or followed by another depolarization phase is called :

(A) Early After Depolarization and Delayed After Depolarization
(B) Conduction block
(C) Cardiac failure
(D) All of the above

190. Which among the following is not a function of vascular endothelium :

(A) Capillary transport and exchange between tissue and blood
(B) Production of clotting factors
(C) Synthesis and secretion of vasodilators and vasoconstrictor substances
(D) Signals the brain regarding painful stimuli

191. Which among the following is not an endothelium derived vasoactive relaxing factor :

(A) Nitrous oxide (B) Prostacyclin
(C) Endothelin (D) Bradykinins

192. Which among the following is false about endothelin (ET) :

(A) It is an endothelium derived vasoactive contraction factor
(B) Made up off 32 amino acids
(C) It has mainly 3 isoforms (D) ET 1 is the potent vasoconstrictor

193. Which among the following is false about smooth muscle contraction :

(A) Like skeletal and cardiac muscle, in smooth muscle actin and myosin filaments are organized
(B) These contraction proteins are scattered in cytoplasm of myocytes
(C) Here the regulatory contractile unit is myosin
(D) There are 2 types of myosin light chain, one pair having ATPase activity and other pair having regulatory activity

194. Myogenic response to stretch is called :

(A) Bayliss mechanism (B) Bezold.Jarish Reflex
(C) Bainbridge reflex (D) Both a and b

195. Which among the following is known as Windkessel vessels :

(A) Aorta (B) Aorta and other big arteries
(C) Aorta and all types of arteries and arterioles
(D) Aorta, big arteries and big veins

196. Pre.capillary resistance vessels are :

(A) Aorta (B) Arteries (C) Arterioles (D) Veules

197. Capacitance vessels include :

(A) Big veins (B) Big veins and venules
(C) Capillaries (D) Big veins, venules and capillaries

198. Cardiac output can be measured by :

(A) The Fick Principle (B) Dye method
(C) Thermal Dilution method (D) All of the above

199. Factors that affect cardiac output includes :

(A) Cardiac inflow
(B) Cardiac inflow and contractibility of heart
(C) Cardiac inflow, contractibility of heart and Heart rate
(D) Cardiac inflow, contractibility of heart, heart rate and coronary blood flow

200. Factors affecting venous return includes :

(A) Cardiac output (B) The respiratory pump activity
(C) The muscle pump and gravity (D) All of the above

201. Heart can generate its own rhythm, this phenomenon is called :

(A) Bayliss mechanism (B) Autorhythmicity
(C) Both the above (D) Cardiac pace

202. Coronary blood flow ceases temporarily during :

(A) Protodiastole (B) End of diastole
(C) Isometric contraction (D) Diastasis

203. Circulation of blood between left auricle and left ventricle is guarded by :

(A) Mirtal valve (B) Tricuspid valve

(C) Aortic valve (D) Pulmonary valve

204. In lead II the electrodes of electrocardiograph are connected to :

(A) Right arm and right leg (B) Left arm and left leg

(C) Right arm and left arm (D) Right arm and left leg

— Answer Sheet —

Q. No.	Ans	Q. No.	Ans	Q. No.	Ans	Q. No.	Ans	Q. No.	Ans	Q. No.	Ans
1.	A	36.	C	71.	C	106.	A	141.	C	176.	C
2.	C	37.	A	72.	B	107.	D	142.	C	177.	B
3.	A	38.	D	73.	C	108.	B	143.	B	178.	A
4.	A	39.	D	74.	A	109.	C	144.	D	179.	B
5.	A	40.	B	75.	A	110.	B	145.	D	180.	B
6.	C	41.	C	76.	C	111.	C	146.	B	181.	B
7.	A	42.	A	77.	B	112.	A	147.	C	182.	A
8.	A	43.	B	78.	D	113.	D	148.	B	183.	B
9.	C	44.	B	79.	A	114.	C	149.	A	184.	D
10.	C	45.	D	80.	A	115.	D	150.	C	185.	A
11.	B	46.	D	81.	B	116.	D	151.	B	186.	A
12.	B	47.	D	82.	D	117.	B	152.	C	187.	C
13.	B	48.	D	83.	C	118.	B	153.	D	188.	A
14.	A	49.	C	84.	C	119.	C	154.	D	189.	A
15.	D	50.	D	85.	C	120.	A	155.	A	190.	D
16.	C	51.	B	86.	A	121.	A	156.	C	191.	C
17.	B	52.	D	87.	C	122.	A	157.	B	192.	B
18.	D	53.	A	88.	B	123.	B	158.	A	193.	A
19.	A	54.	D	89.	C	124.	A	159.	D	194.	A
20.	B	55.	D	90.	B	125.	B	160.	D	195.	B
21.	A	56.	C	91.	D	126.	C	161.	B	196.	C
22.	D	57.	B	92.	A	127.	C	162.	A	197.	A
23.	A	58.	A	93.	C	128.	D	163.	A	198.	D
24.	C	59.	B	94.	B	129.	B	164.	C	199.	C
25.	A	60.	B	95.	D	130.	A	165.	A	200.	D
26.	C	61.	A	96.	C	131.	A	166.	A	201.	B
27.	A	62.	C	97.	D	132.	C	167.	C	202.	C
28.	B	63.	D	98.	D	133.	A	168.	A	203.	A
29.	A	64.	D	99.	D	134.	C	169.	D	204.	D
30.	A	65.	C	100.	C	135.	D	170.	D		
31.	C	66.	A	101.	D	136.	C	171.	A		
32.	D	67.	B	102.	B	137.	C	172.	A		
33.	B	68.	C	103.	B	138.	C	173.	B		
34.	D	69.	C	104.	B	139.	A	174.	C		
35.	B	70.	B	105.	C	140.	B	175.	D		

• Chapter 3 •

Body Fluids and Cell Physiology

1. **Animal body contains .... percent water.**
   (A) 5-10% (B) 15-20% (C) 25-40% (D) 50-60%

2. **1 mole of NaCl is equal to .... gram.**
   (A) 58g (B) 67g (C) 58.5g (D) 67.5g

3. **Which among the following is not a colligative property :**
   (A) Vapor pressure lowering (B) Boiling point elevation
   (C) Freezing point depression (D) Diffusion

4. **In which of these compartments is Na^+ concentration the lowest :**
   (A) Interstitial fluid (B) Intracellular fluid
   (C) Lymph (D) Plasma

5. **Buffer systems do which of the following :**
   (A) minimize pH changes in body fluids by converting stronger acids or bases into weaker acids or bases
   (B) consist of a weak acid plus a strong acid
   (C) are not present in intracellular fluid
   (D) increase the severity of pH changes in the body fluids

6. **Body water is lost through each of these organs EXCEPT the :**
   (A) Kidney (B) Muscle
   (C) Lungs (D) Skin

7. **A loss of body water would do which of the following :**
   (A) inhibit the thirst center
   (B) stimulate osmoreceptors in the hypothalamus
   (C) inhibits ADH secretion
   (D) all the above

8. **Which condition results water output to exceed water intake :**
   (A) water intoxication (B) dehydration
   (C) increase blood volume (D) ADH hyposecretion

9. **A diuretic is which of the following :**
   (A) a drug that may be administered in order to reduce water loss through sweating
   (B) a drug that may be administered in order to prevent water excretion in the urine
   (C) a substance that causes increased urine production
   (D) a substance that causes urine retention

10. **The hormone directly increases water reabsorption in the kidneys :**
    (A) Aldosterone (B) Atrial natriuretic peptide
    (C) Vasopressin (D) Insulin

11. **Which factor stimulates ADH secretion :**
    (A) high ECF volume (B) high ECF osmolarity
    (C) high blood pressure (D) high plasma calcium

12. A small amount of the daily water requirement is met by molecules produced during oxidative phosphorylation in the mitochondria. This is referred :

(A) metabolic water (B) transcellular water
(C) cutaneous water (D) free water

13. Which factor would cause a shift of water from the plasma to the interstitial fluid :

(A) increased interstitial fluid hydrostatic pressure
(B) increased blood plasma colloid osmotic pressure
(C) decreased blood pressure (D) increased blood pressure

14. These are primarily responsible for the movement of fluid between plasma and interstitial fluids :

(A) primary and secondary active transport
(B) membrane transport
(C) diffusion and osmosis
(D) hydrostatic and osmotic pressure

15. The interstitial fluid, plasma and lymph are part of which larger fluid compartment :

(A) Intracellular fluid (B) Transcellular fluid
(C) Extracellular fluid (D) All of the above

16. The major anion in the intracellular fluid is :

(A) Phosphate (B) Sulphate (C) Sodium (D) Chloride

17. The major anion in the intracellular fluid is :

(A) Phosphate (B) Sulphate (C) Sodium (D) Chloride

18. In the kidney tubules, aldosterone increases reabsorption of .... but decreases reabsorption of ....

(A) K^+, Na^+ (B) Na^+, K^+ (C) Ca^{2+}, PO_4 (D) PO_4, Ca^{2+}

19. The major cation in the intracellular fluid is :

(A) Sodium (B) Potassium (C) Calcium (D) Magnesium

20. The major cation in the extracellular fluid is :

(A) Sodium (B) Potassium (C) Calcium (D) Magnesium

21. Strong acids :

(A) ionize less completely than weak acids
(B) can be buffered by weak bases
(C) cause a smaller change in pH than weak acids
(D) release fewer H^+ than weak acids

22. Which of these chemicals is capable of buffering strong bases :

(A) $NaHCO_3$ (B) H_2CO_3 (C) NaCl (D) H_2PO_4

23. Prolonged diarrhea can cause :

(A) Metabolic acidosis (B) Respiratory acidosis
(C) Metabolic alkalosis (D) Respiratory acidosis

24. Prolonged vomiting may cause :

(A) Metabolic acidosis (B) Respiratory acidosis
(C) Metabolic alkalosis (D) Respiratory alkalosis

25. Hyperventilation results :
(A) Metabolic acidosis (B) Respiratory acidosis
(C) Metabolic alkalosis (D) Respiratory alkalosis

26. Hypoventilation results :
(A) Metabolic acidosis (B) Respiratory acidosis
(C) Metabolic alkalosis (D) Respiratory alkalosis

27. Kidney disease causes :
(A) Metabolic acidosis (B) Respiratory acidosis
(C) Metabolic alkalosis (D) Respiratory acidosis

28. The normal pH of the blood ranges between :
(A) 6-7 (B) 7-8 (C) 7.35-7.45 (D) 6.98-7.24

29. As the hydrogen ion concentration of the ECF increases, the :
(A) PH of ECF increases
(B) Kidney increase hydrogen ion secretion
(C) PH of urine increases
(D) Respiratory rate decreases

30. Which ite.n is correctly matched with the area where it is located or secreted :
(A) Aldosterone: adrenal medulla
(B) ADH secretion: anterior pituitary
(C) Thirst center: hypothalamus
(D) Atrialnatriuretic peptide: adrenal medulla

31. Cerebrospinal fluid, fluid within the eyes, joints, and body cavities, and fluid secretions of exocrine glands are all classified specifically as .... fluid.
(A) Transcellular fluid (B) Interstitial fluid
(C) Extracellular fluid (D) All are true

32. What is the main force that causes water to move among the various fluid compartments?
(A) Osmosis (B) Filtration
(C) Bulk movement (D) Hydrolytic pressure

33. Which of these chemicals is capable of buffering strong bases :
(A) $NaHCO_3$ (B) NaCl (C) H_2CO_3 (D) HPO_4

34. Which of these body fluid compartments contains the highest concentration of protein :
(A) Intracellular (B) Interstitial
(C) Plasma (D) Cerebrospinal fluid

35. In which choice are the major body fluid compartments ranked correctly in order of decreasing volume :
(A) Interstitial, intracellular, plasma (B) Intracellular, Interstitial, plasma
(C) Plasma, Interstitial, intracellular (D) Intracellular, plasma, Interstitial

36. The fluid secreted into the stomach by the stomach lining is part of which fluid compartment?
(A) Interstitial (B) Extracellular
(C) Transcellular (D) Plasma

37. The capillary endothelium separates these two body fluid compartments :
(A) ICF and ECF
(B) ICF and plasma
(C) Interstitial fluid and Plasma
(D) CSF and interstitial fluid

38. The main buffer in the interstitium is :
(A) protein (B) phosphate (C) ammonia (D) HCO_3

39. Regarding the function of the smooth endoplasmic reticulum; which is incorrect :
(A) Steroid synthesis
(B) Drug detoxification
(C) Protein synthesis
(D) Role in carbohydrate metabolism

40. Evan's blue dye is used to measure the .... volume.
(A) Total body water
(B) Plasma
(C) Extracellular fluid
(D) Intracellular fluid

41. Which compartment of body fluid cannot be measured directly by tracers :
(A) Interstitial fluid
(B) Intracellular fluid
(C) Transcellular fluid
(D) All of the above

42. Antipyrine is used to measure :
(A) Total body water
(B) Plasma
(C) Interstitial fluid
(D) All of the above

43. Mitrocondria have been first seen by :
(A) Robert Hooke
(B) Robert Brown
(C) Lipmann
(D) Altmann

44. Plasma membrane is composed of :
(A) Protein (B) Lipid (C) Cellulose (D) Protein and lipid

45. The first successfully cloned animal was :
(A) Monkey (B) Rabbit (C) Sheep (D) Buffalo

46. Which statement is false about membrane transport :
(A) Carrier mediated transport is slower than channel mediated transport
(B) Only small nonpolar substances can cross membrane by simple diffusion
(C) Only large nonpolar molecules are carried via channel mediated transport
(D) Large polar molecules are carried via carrier mediated transport

47. As lysosome originate from Golgi apparatus, peroxisome is from :
(A) Rough endoplasmic reticulum
(B) Smooth endoplasmic reticulum
(C) Similar to lysosome, i.e. from Golgi apparatus
(D) Ribosome

48. PH of lysosome is closer to :
(A) 6.8 – 7.4 (B) 7 – 8 (C) 5 (D) 6 – 8

49. Blood brain barrier was discovered by :
(A) Paul Ehrlich
(B) Metchnikoff
(C) Robert Brown
(D) Lipmann

50. Metabolically active cell has more numbers of :
(A) Nucleolus
(B) Ribosome
(C) Golgi apparatus
(D) Mitochondria

51. Which one is not perfectly matched :
(A) Tight junction ...... Claudin, Occludin
(B) Gap junction .... Connexon
(C) Desmosome ....Claherin
(D) All are true

52. Which one correctly defines one Molal solution :
(A) Number of moles of substance in 1 liter water/solvent
(B) Number of moles of substance in 1 Kg water/solvent
(C) Number of grams of substance in 1 liter water/solvent
(D) Number of grams of substance in 1 Kg water/solvent

53. Selectively permeable membrane :
(A) Only permeable to water
(B) Permeable to water along with some ions
(C) Only permeable to some ions
(D) Permeable to proteins

54. Effective osmotic pressure is due to :
(A) Solute that cannot pass through the membrane
(B) Solute that can pass through the membrane
(C) Water and solute that pass through the membrane
(D) Water only

55. Sweating and panting are .... type of water loss.
(A) Sensible water loss (B) Insensible water loss
(C) Non.evaporative water loss (D) None of these

56. Sweat gland exhaustion after profuse sweating is called :
(A) Estivation (B) Hidrosis
(C) Anhidrosis (D) Exhaustive hidrosis

57. Metabolic water in most domestic animals accounts nearly .... percent of body weight.
(A) 5-10 (B) 20-40 (C) 40-60 (D) 60-70

58. Water intoxication is
(A) Cellular overhydration (B) Intravascular overhydration
(C) Interstitial overhydration (D) Excessive water loss

59. When there is dehydration the immediate source of water loss is from :
(A) ECF (extracellular fluid)
(B) ICF (intra cellular fluid)
(C) From all compartments with equal proportion
(D) Lymph

60. Sheep can tolerate dehydration of about .... percent of their body weight :
(A) 10 – 15 (B) 15 – 20
(C) 20 – 25 (D) 25 – 30

61. Unequal distribution of diffusible ions across the capillary wall is called :
(A) Vant Hoff's phenomena (B) Starling's phenomena
(C) Gibb Donnan's phenomena (D) Hamberger's phenomena

62. Gibb Donnan's phenomena is due to :
(A) Diffusibleions (B) Proteins
(C) Water (D) Cations

63. Reflection coefficient (σ) is a correction factor used to calculate :
(A) The effective osmotic pressure (B) The ion flow
(C) The diffusion coefficient (D) The bulk flow

64. Which one is false about tonicity :
(A) Also called effective osmotic pressure
(B) Can also be called osmolarity
(C) It is a property of membrane as well as solution
(D) Osmolarity is only particle property

65. Which one is false about Na^+ K^+ ATPase pump :
(A) It is a carrier protein
(B) It has one α and one β subunit
(C) It carries 3 Na^+ and 2 K^+ ion
(D) It has no role in maintaining the membrane in polarized state

66. Blood brain barrier (BBB) is not permeable to :
(A) Sodium ion (B) Oxygen (C) Carbon dioxide (D) Glucose

67. Matured RBC lacks
(A) Nucleus (B) Endoplasmic reticulum
(C) Golgi apparatus (D) Both (A) and (B)

68. Cells synthesizing steroids have a higher number of :
(A) Rough endoplasmic reticulum (B) Smooth endoplasmic reticulum
(C) Golgi apparatus (D) Mitochondria

69. Lysosomes are produced from :
(A) Rough endoplasmic reticulum (B) Smooth endoplasmic reticulum
(C) Golgi apparatus (D) Mitochondria

70. Proteins synthesized by the rough ER are :
(A) for internal storage (B) for internal regulation
(C) to digest food in lysosome (D) exported from the cell

71. Clusters of rRNA where ribosomes are assembled are called :
(A) Nuclei (B) Centriole (C) Nucleolus (D) Cisternae

72. The smooth ER is especially abundant in cells that synthesize extensive amounts of :
(A) Toxin (B) Protein (C) Enzyme (D) Lipid

73. The cytoskeleton includes all of the following except :
(A) Microtubules (B) Intermediate filament
(C) Myosin filament (D) Actin filament

74. Which of the following is not true of Golgi apparatus :
(A) They were first observed by Camillo Golgi
(B) Golgi apparatus is found in Prokaryotic cells
(C) Cis Phase of cisternae is the forming face
(D) The main function is to package materials for transport

75. Size of the ribosomes is indirectly indicated by 'S' which stands for :
(A) Subunit (B) Sedimentation coefficient
(C) Shape (D) Size

76. Larger molecules like glucose attach to a carrier molecule that aids their transport through the membrane, is called :
(A) Diffusion (B) Facilitated diffusion
(C) Active transport (D) Osmosis

77. The plasma concentration of sodium is regulated by :
(A) Potassium (B) Chloride
(C) Aldosterone (D) Thirst-ADH system

78. The cell organelle that increases its number by self-replication :
(A) Endoplasmic reticulum (B) Nucleolus
(C) Mitochondria (D) Nuclei

79. The intracellular connection having high perneability of ions is :
(A) Desmosome (B) Hemidesmosome
(C) Focal adhesion (D) Gap junction

80. Facilitated diffusion differs from simple diffusion by :
(A) It involves a carrier protein (B) Is an active transport
(C) Is a passive transport (D) Occurs due to electrical gradient

81. PH of extracellular fluid seldom varies from :
(A) 7.0 (B) 7.4 (C) 7.6 (D) 8.0

82. PH= PKa$^+$ log [base/acid] :
(A) Henderson–Hasselbalch equation
(B) Hydration equation
(C) Dissociation reaction
(D) Both (A) and (B)

83. A chemical buffer system includes :
(A) Weak base and strong acid (B) Strong base and weak acid
(C) Weak acid and its conjugate base (D) Both (A) and (B)

84. PKa of bicarbonate buffer is :
(A) 7 (B) 6.1 (C) 6.8 (D) 7.4

85. Most important chemical buffer system of our body is :
(A) Phosphate buffer (B) Bicarbonate buffer
(C) Ammonia buffer (D) All of the above

86. PKa of bicarbonate buffer is :
(A) 7 (B) 6.1 (C) 6.8 (D) 7.4

87. .... amino acid is present in haemoglobin.
(A) Aspartate (B) Histidine
(C) Tryptophan (D) Lysine

88. Most important buffer in intra.cellular fluid is :
(A) Phosphate buffer (B) Bicarbonate buffer
(C) Ammonia buffer (D) All of the above

89. Phosphate buffer is important in :
(A) Extra-cellular fluid (B) Intra-cellular fluid
(C) Tubular fluid of Kidney (D) Both (B) and (C)

90. Isohydric principle is :
(A) Buffers of blood and body fluid interact in unison
(B) Venous blood is more basic than arterial blood
(C) The reversible reaction occurring in chemical buffering system
(D) None of the above

91. Hypokalemia indicates :
(A) Decreased concentration of plasma sodium concentration
(B) Decreased concentration of plasma potassium concentration
(C) Decreased concentration of plasma hydrogen concentration
(D) Both (A) and (B)

92. Normal intracellular potassium concentration is :
(A) 4 mEq/L (B) 50 mEq/L
(C) 140 mEq/L (D) 1500 mEq/L

93. GLUT-1 is a specific glucose carrier is an example of :
(A) Simple diffusion (B) Facilitated diffusion
(C) Active transport (D) Both (B) and (C)

94. Glucose in CNS is carried by :
(A) GLUT-1 (B) Simple diffusion
(C) Na^+ dependent symport carrier (D) Na^+/K^+ ATPase

95. Water comprises ....% of total body weight in average adult animals.
(A) 20 (B) 40 (C) 60 (D) 80

96. Simple diffusion is under the influence of :
(A) Brownian motion (B) Concentration gradient
(C) Against concentration gradient (D) Both (A) and (B)

97. Semipermeable membrane is :
(A) Permeable to water but not to solutes
(B) Permeable to water and some solutes but impermeable other solutes
(C) Permeable to all solutes in the direction of concentration gradient
(D) All are true

98. Osmosis is :
(A) Movement of water (B) Down the concentration gradient
(C) Needs semipermeable membrane (D) All are true

99. The membranes of body are :
(A) Completely permeable (B) Semipermeable
(C) Selectively permeable (D) Can be any one of the above

100. Effective osmotic pressure is mainly contributed by :
(A) Water
(B) Solutes that are permeable through membrane
(C) Solutes that are impermeable through membrane
(D) Both (A) and (B)

101. Crenation results when cells are kept in :
(A) Hypertonic solution (B) Hypotonic solution
(C) Isotonic solution (D) Any one of (A) and (B)

102. The erythrocyte of which species are more resistance to hemolysis when kept in hypotonic solution :
(A) Dog (B) Goat (C) Sheep (D) Both (B) and (C)

103. Out of total body water (60%) which compartment contains highest :
(A) Extracellular fluid (40%) (B) Intracellular fluid (40%)
(C) Interstitial fluid (40%) (D) Plasma (40%)

104. Fluid found in different body cavities is called :
(A) Extracellular fluid (B) Intracellular fluid
(C) Inter cellular fluid (D) Trans-cellular fluid

105. Water intoxication most appropriately describes as :
(A) Cellular overhydration (B) ECF expansion
(C) Plasma volume expands (D) Anyone of the above can occur

106. Vapor losses are called as :
(A) Insensible water loss (B) Sensible water loss
(C) Respiratory water loss (D) Both (A) and (B)

107. Loss of water from extra cellular fluid(ECF) is called :
(A) Water scarcity (B) Hypertonicity (C) Dehydration (D) Hydration

108. Which one of the following sentence is correct relative to water tolerance :
(A) Cattle have better tolerance than sheep
(B) Sheep, cattle and pig have same tolerance
(C) Pigs are better tolerance than sheep
(D) Sheep have better tolerance than cattle and pig

109. What are the ions that are used in calculation of anion gap :
(A) Na^+, K^+, Mg_2^+ and Ca_2^+ (B) Na^+, K^+, Cl^- and SO_4^-
(C) Na^+, K^+, Cl^- and HCO_3^- (D) Cl^-, SO_4^-, HCO_3^- and PO_4^-

110. In normal case anion gap varies between :
(A) 2-20 mEq/L (B) 8-16 mEq/L
(C) 10-30 mEq/L (D) 20-40 mEq/L

111. In metabolic acidosis :
(A) Anion gap decreases (B) Anion gap Increases
(C) Anion gap unaltered (D) Not fixed anyone can occur

112. Anion gap is mainly helpful in :
(A) Hypoproteinemic alkalosis (B) Metabolic acidosis
(C) Both (A) and (B) (D) Metabolic and Respiratory acidosis

113. The typical reference range of base excess varies between :
(A) –2 to +2 mEq/L (B) –6 to +6 mEq/L
(C) –8 to +8 mEq/L (D) –10 to +10 mEq/L

114. Predominant base contributes in base excess is :
(A) Chloride (B) Bicarbonate (C) Phosphate (D) Sulphate

115. More than half of total protein in mammalian body is :
(A) Histones (B) Albumins (C) Collagen (D) Haemoglobin

116. Secretory granules are synthesized from :
(A) Golgi apparatus (B) Smooth Endoplasmic Reticulum
(C) Rough Endoplasmic Reticulum (D) Ribosome

117. Glycosylation of proteins and peptides occurs at :
(A) Golgi apparatus (B) Smooth Endoplasmic Reticulum
(C) Rough Endoplasmic Reticulum (D) Ribosome

118. Transport vesicles pinch off from :
(A) Ribosome (B) Endoplasmic Reticulum
(C) Golgi apparatus (D) All of the above

119. Peptidase in lysosome is :
(A) Lysozyme (B) Cathepsin
(C) Pepsinase (D) All of the above

120. By Fick's law of diffusion rate of diffusion is directly proportional to :
(A) Surface Area (B) Diffusion distance
(C) Pressure difference (D) Both (A) and (C)

121. Non diffusible ions trapped in a membrane attract opposite charges and repel similar ones is explained in :
(A) Fick's law (B) Gibbs-Donnan effect
(C) Starling's Hypothesis (D) Van't Hoff's Law

122. Which one is true about Reflection Coefficient :
(A) Used as a correction factor to find out effective osmotic pressure
(B) Designated as σ
(C) When $\sigma = 0$: membrane is impermeable to both water and solute, when $\sigma = 1$: membrane shows complete reflection to solute and is completely permeable to water
(D) All of the above are true

123. Which statement accurately describes Poiseuille's law :
(A) It explains rate of diffusion (B) It explains fluid movement
(C) It explains bulk flow (D) It explains tonicity

124. The property that is measured by the number of particles but not the mass of particles is called :
(A) Osmolar property (B) Tonicity
(C) Colligative property (D) Reflective property

125. Which among the following is not included in colligative property :
(A) Freezing point depression (B) Vapor pressure depression
(C) Size of the particle (D) Osmotic pressure

126. Which among the following is true about channel transport :
(A) Allow passage of polar and small particles
(B) Exhibit Vmax
(C) Transport protein exhibit conformation change or flip
(D) Can carry only one molecule at a time, so requires more time

127. Which statement is false about Na^+ K^+ ATPase pump :
(A) It is a carrier protein with one α and one β subunit
(B) Inner side has ability to bind 3 Na^+ and one ATP binding site
(C) Outer side has ability to bind 2 K^+
(D) Keeps membrane in non-polarised state

128. Which among the following is true about cell to cell junction :
(A) Present towards the apical surface
(B) It is made with protein Claherine
(C) Allow passage of ion, which make cells electrically connected
(D) Two cells are separated at a distance of 200 Å

129. What is the role of a biological membrane :
(A) Compartmentalization
(B) Selective transport
(C) Receptor mediated signal transduction
(D) All of the above

130. Which one is incorrectly matched :
(A) Tight junction – Claudine and Occludin
(B) Gap junction – Connexon
(C) Desmosome – Claherine
(D) Desmosome – Claudine

131. Ribosomes are synthesized by :
(A) Nucleolus (B) Golgi
(C) Mitrochondria (D) Lysosome

132. Protein present in chromatin is mainly :
(A) Collagen (B) Histone (C) Non-histone (D) Both (B) and (C)

133. RER (Rough Endoplasmic Reticulum) number heavily outnumber SER (Smooth Endoplasmic Reticulum) in :
(A) Liver cell (B) Adrenal cortical cell
(C) Both (A) and (B) (D) All type of cells

134. Proteins synthesized by free ribosomes :
(A) Are released in to cytosol (B) Are essential for cell it self
(C) Both (A) and (B)
(D) Are transported to Golgi via Endoplasmic Reticulum

135. Which one among the following is true about lysosome :
(A) These are unit membrane bound organelle
(B) They contain acid hydrolases
(C) The pH within the lysosome is acidic
(D) All of these are true

136. Which among the following is false about cytoskeletons :
(A) Includes microtubules, microfilaments and intermediary filaments
(B) Microtubules are made up off a protein called Tubulin and includes cilia, spindles and processes of neurons
(C) Microfilament is mainly actin
(D) Microfilament is the main element in centriole

137. Most cells contain 70-85 % water except :
(A) Liver cell (B) Fat cell (C) Muscle cell (D) All of these

138. After water the most abundant substance in cell is :
(A) Protein (B) Fat (C) Carbohydrate (D) Ash

139. Carrier proteins present in cell membrane are :
(A) Integral protein
(B) Peripheral proteins
(C) Proteins protruding only inside of the cell
(D) Both (A) and (B)

140. Peroxisomes differ from lysosome :
(A) Peroxisomes are formed by self-replication
(B) It comes out from ER but not from Golgi
(C) It contains oxidase rather than hydrolases
(D) All are true

— Answer Sheet —

Q. No.	Ans	Q. No.	Ans	Q. No.	Ans	Q. No.	Ans	Q. No.	Ans	Q. No.	Ans
1.	D	26.	B	51.	D	76.	B	101.	A	126.	A
2.	C	27.	C	52.	B	77.	D	102.	A	127.	D
3.	D	28.	C	53.	B	78.	C	103.	B	128.	A
4.	B	29.	B	54.	A	79.	D	104.	D	129.	D
5.	A	30.	C	55.	B	80.	A	105.	A	130.	D
6.	B	31.	A	56.	C	81.	B	106.	A	131.	A
7.	B	32.	A	57.	A	82.	A	107.	C	132.	B
8.	D	33.	C	58.	A	83.	C	108.	D	133.	A
9.	C	34.	A	59.	A	84.	B	109.	C	134.	C
10.	C	35.	D	60.	D	85.	B	110.	B	135.	D
11.	B	36.	C	61.	C	86.	C	111.	B	136.	D
12.	A	37.	C	62.	B	87.	B	112.	C	137.	B
13.	D	38.	D	63.	A	88.	A	113.	A	138.	A
14.	D	39.	C	64.	B	89.	D	114.	B	139.	A
15.	C	40.	B	65.	D	90.	A	115.	C	140.	D
16.	A	41.	D	66.	A	91.	B	116.	A		
17.	D	42.	A	67.	D	92.	C	117.	A		
18.	B	43.	D	68.	B	93.	B	118.	B		
19.	B	44.	D	69.	C	94.	A	119.	B		
20.	A	45.	C	70.	D	95.	C	120.	D		
21.	B	46.	C	71.	C	96.	D	121.	B		
22.	A	47.	B	72.	D	97.	A	122.	D		
23.	A	48.	C	73.	C	98.	D	123.	C		
24.	C	49.	A	74.	B	99.	C	124.	C		
25.	D	50.	D	75.	B	100.	C	125.	C		

• Chapter 4 •
Digestive System

1. **End product of fermentation in ruminants is :**
(A) Glucose (B) Fatty acid
(C) Volatile fatty acid (D) Both (A) and (C)

2. **Which part of ruminant stomach is kidney shaped :**
(A) Rumen (B) Reticulum (C) Omasum (D) Abomasum

3. **In abomasum mucosa is thrown into 12 high folds called as :**
(A) Club shaped papillae (B) Rugae
(C) Lamella (D) Leaves

4. **Which among the following is true about tension receptors :**
(A) Slow adapting mechanoreceptors
(B) Readily adapting mechanoreceptors
(C) Readily adapting chemoreceptors
(D) Lie close to basement membrane of luminal epithelium

5. **Which among the following is a pseudoruminant :**
(A) Cattle (B) Sheep (C) Pig (D) Camel

6. **Which among the following is true about ruminal ecosystem :**
(A) Ruminal environment is strictly anaerobic
(B) Ruminal PH is 6.8
(C) House of billions of microbes
(D) All are true

7. **Which one is false about cellulytic bacteria :**
(A) Optimally act as PH 6.2
(B) They need amino acid for protein synthesis
(C) Fermentation is slow, with doubling time of about 18 hours
(D) Acetate is the major VFA produced by them

8. **Which one is false about amylolytic bacteria :**
(A) Optimally act as PH 6.2
(B) They need amino acid and ammonia for protein synthesis
(C) Fermentation is fast, with doubling time of about 0.5-2.5 hours
(D) Propionate is the major VFA produced by them

9. **About .... liter gas is produced by rumen in one hour after 2.3 hr. of feeding.**
(A) 20 lit/hr. (B) 40 lit/hr. (C) 50 lit/hr. (D) 60 lit/hr.

10. **Which gas is produced in rumen maximally :**
(A) CO_2 (B) CH_4 (C) NH_3 (D) H_2

11. **.............. amount of saliva is produced by cattle per day.**
(A) 8-16 L/day (B) 10-100 L/day
(C) 16-160 L/day (D) 20-200 L/day

12. **PH of cow saliva is about :**
(A) 6.8 (B) 7.4 (C) 8.1 (D) 8.9

13. Which statement is false about major salivary gland secretion in cattle :
(A) Secretion is mucous in nature
(B) Highly buffered
(C) Isotonic
(D) Secretion is by stimulation from mouth, esophagus and rumino-reticulum

14. Salivary center is present at :
(A) Forebrain (B) Mid brain (C) Hind brain (D) Cerebellum

15. Particles for rumination should be of .... size.
(A) <3 mm (B) >3 mm ..... <10 mm
(C) >10 mm.....<300 mm (D) >300 mm...... <1000 mm

16. Primary cycle in rumino.reticular motility is associated with :
(A) Mixing (B) Segregation (C) Eructation (D) Both (A) and (B)

17. Secondary cycle in rumino-reticular motility is associated with :
(A) Mixing (B) Segregation (C) Eructation (D) Both (A) and (B)

18. Except .... cells all other cells migrate from base of crypt to tip.
(A) Chief cell (B) Mucous neck cells
(C) Paneth cells (D) Crypt epithelial cells

19. VFA provides .... percent of total energy need of ruminants :
(A) 10-20 (B) 20-40 (C) 40-60 (D) 60-80

20. In randomization pathway, fate of 2 pyruvate is :
(A) 1 propionate + 1 acetate (B) 2 Propionate
(C) 2 acetate (D) 1 butyrate

21. 1 mole of ATP gives about .... gram of bacterial dry matter.
(A) 10-20 gram (B) 20-30 gram
(C) 15-45 gram (D) 20-40 gram

22. Protozoa population in rumen liquor is about :
(A) 10^3 - 10^4 cells /gram of rumen liquor
(B) 10^5 - 10^6 cells /gram of rumen liquor
(C) 10^{10} - 10^{11} cells /gram of rumen liquor
(D) 10^{11} - 10^{12} cells /gram of rumen liquor

23. Bacteria population in rumen liquor is about :
(A) 10^3 - 10^4 cells /gram of rumen liquor
(B) 10^5 - 10^6 cells /gram of rumen liquor
(C) 10^{10} - 10^{11} cells /gram of rumen liquor
(D) 10^{11} - 10^{12} cells /gram of rumen liquor

24. Which one is true about fermentative digestion :
(A) Rate of reaction is fast
(B) Extend of alteration is to a great extent
(C) Enzymes are from host
(D) All are true

25. Which one is true about avian digestion :
(A) Teeth is absent (B) Hard palate is absent
(C) Muscular stomach is absent (D) All are true

26. Which one is true about crop :
(A) It is a dilatation of esophagus (B) Absent in owl
(C) Two pouched in pigeon (D) All are true

27. Which one is false about gizzard :
(A) Is muscular stomach in birds
(B) Has 2 pairs of muscle
(C) Mucosal membrane is covered by thick koilin membrane
(D) Nerve of Remak is supplied to it

28. Salivary glands in birds produce :
(A) about 7.30 ml saliva /day (B) mucinous saliva
(C) amylase containing saliva (D) all are true

29. PH of proventriculus is :
(A) 0.5-2.5 (B) 1.8-3.9 (C) 3.8-6.8 (D) 8-8

30. Gastric juice produced in birds is about :
(A) 0.5ml/hr./Kg body weight (B) 3.5ml/hr./Kg body weight
(C) 6.8ml/hr./Kg body weight (D) 8.8ml/hr./Kg body weight

31. Intestinal PH I birds is :
(A) 4.5-5.8 (B) 5.6-7.2 (C) 7-8 (D) 8-9

32. In monogastric digestion where is bacteria present that is needed for digestion:
(A) Esophagus (B) Caeca (C) Cecum (D) Stomach

33. Bile is produced from :
(A) Liver (B) Stomach (C) Kidney (D) Gall bladder

34. Brunner's gland is found in :
(A) Stomach (B) Duodenum (C) Jejunum (D) Ileum

35. Duodenal hormone helping in neutralization of acidic chyme of stomach :
(A) Secretin (B) Cholecystokinin
(C) Entero-pancreozymin (D) Both (A) and (B)

36. Which one is false :
(A) Ruminants digest cellulose more extensively than any other animal species.
(B) All ruminants are pregastric fermenters
(C) Forestomachs include the abomasum also
(D) Goats and sheep able to be more selective than cows in their grazing habits

37. Which among the following is not a ruminant :
(A) Cow (B) Giraffe (C) Elephant (D) Deer

38. Why are goats and sheep able to be more selective than cows in their grazing habits?
(A) They have mobile lip (B) They have a narrow muzzle width
(C) They have a nimbler tongue (D) Both (A) and (B)

39. The only gut structure which has striated muscle in most domestic animal is :
(A) Esophagus (B) Stomach (C) Duodenum (D) Colon

40. The major muscular reflex in proximal stomach is :
(A) Adaptive relaxation (B) Contraction
(C) Grinding (D) Sieving

41. Through pyloric sphincter particles of size .... can only pass to duodenum.
(A) 2 mm (B) 3 to 6 mm (C) 7 to 10 mm (D) 11 to 20 mm

42. Out of the following gastric hormone which one facilitates gastric motility :
(A) CCK (B) Gastrin (C) GIP (D) Secretin

43. Which one of the following is a potent stimuli for entero-gastric reflex :
(A) Low PH (B) High Osmolarity
(C) Presence of fat (D) All of the above

44. Vomition center is located at :
(A) Hypothalamus (B) Thalamus
(C) Median eminence (D) Brain stem

45. Which one of the following is a hindgut fermenter :
(A) Cattle (B) Horse (C) Deer (D) Giraffe

46. Kreb's cycle occurs in all tissue except in :
(A) Neurons (B) Hepatocytes (C) RBC (D) Adipose tissue

47. In ruminants ketone bodies are formed in :
(A) Liver (B) Ruminal epithelium
(C) Both (A) and (B) (D) In all organs requiring energy

48. Amount of glycogen stored in liver does not exceed .... percent of liver weight.
(A) 2 (B) 5 (C) 10 (D) 20

49. Some metabolites enter pointer portal circulation after their synthesis in gut but cannot enter general circulation as they get degraded in liver this phenomena is called :
(A) Hepatic degradation (B) Early phase response
(C) First pass effect (D) All of the above

50. .... has a stimulatory effect on insulin and glucagon production.
(A) Glucose (B) Carbohydrate other than glucose
(C) Fat (D) Protein

51. The saliva helps in digestion of :
(A) Protein (B) Fat (C) Fiber (D) Starch

52. Kuffer's cells are found in :
(A) Brain (B) Liver (C) Kidney (D) Intestine

53. Which of the following mammalian cell type is not capable of metabolizing glucose to carbon dioxide aerobically :
(A) Red blood cells (B) White blood cells
(C) Unstriated muscle cells (D) Liver cells

54. Match the organ of prehension with species :

(A) Flexible upper and lower lips	a. Horse
(B) Long flexible tongue	b. Pig
(C) Teeth	c. Cattle
(D) Mobile upper lip	d. Dog

(A) A–a, B–b, C–c, D–d (B) A–a, B–c, C–d, D–b
(C) A–b, B–a, C–d, D–c (D) A–b, B–a, C–d, D–c

55. Which one is odd among the following functionally :
(A) GIP (B) VIP (C) Glucagon (D) Gastrin

56. What phase of gastric secretion absent in herbivorous :
(A) Cephalic phase (B) Gastric phase
(C) Cardiac phase (D) None of these

57. 75% of gastric juice is secreted in which phase :
(A) Cephalic phase (B) Gastric phase
(C) Cardiac phase (D) None of these

58. In ruminants parital cell secrets which of these :
(A) HCl (B) Intrinsic factor
(C) Both (A) & (B) (D) Pepsinogen

59. Chief cells in ruminant secrets :
(A) HCl (B) Intrinsic factor
(C) Both (A) & (B) (D) Pepsinogen

60. Gastric juice concentration is regulated by :
(A) Feedback mechanism of Duodenum
(B) Hormonal factor of duodenum
(C) Direct means
(D) All of the above

61. pH of small intestine is maintained by :
(A) HCO_3 (B) HCl (C) HNO_3 (D) HCL and HNO_3

62. pH of small intestine is :
(A) 1-2 (B) 2-3 (C) 6.5-7.0 (D) 8-9

63. Concentration of volatile fatty acid in rumen is :
(A) 100-150 mm (B) 60 mm (C) 70 mm (D) 50 mm

64. Trypsinogen activated to trypsin by which of the following :
(A) Enterokinase (B) HCl (C) Trypsin (D) $NaHCO_3$

65. Chymotrypsinogen is activated to chymotrypsin by which of the following :
(A) Enterokinase (B) HCl (C) Trypsin (D) $NaHCO_3$

66. Pancreatic digestive enzymes are called as :
(A) Insulin (B) Glucagon
(C) Both (A) and (B) (D) Zymogen

67. Odour of ruminal content in normal state is :
(A) Ammonical (B) Vinegar like (C) Fishy putrid (D) Sour

68. In acidosis ruminal content smell like :
(A) Ammonical (B) Vinegar like (C) Fishy putrid (D) Sour

69. In acidosis ruminal content smell like :
(A) Ammonical (B) Vinegar like (C) Fishy putrid (D) Sour

70. Percent of methane (CH_4) in rumen is :
(A) 6 (B) 7 (C) 25 (D) 65

71. Percent of carbon dioxide (CO_2) in rumen is :
(A) 6 (B) 7 (C) 25 (D) 65

72. Percent of Nitrogen in rumen is :
(A) 6 (B) 7 (C) 25 (D) 65

73. Percent of Oxygen in rumen is :
(A) 6 (B) 7 (C) 25 (D) 65

74. Which is the diverticulum of oesophagus in birds :
(A) Gizzard (B) Duodenum (C) Crop (D) Stomach

75. Which of the following is the site of fermentation in birds :
(A) Crop (B) Caecum
(C) Both the above (D) Gizzard

76. Glandular stomach is known as :
(A) Gizzard (B) Crop
(C) Proventriculus (D) All of the above

77. Gizzard is mainly for :
(A) Crushing (B) Chewing (C) Digestion (D) None of these

78. Which of the following organs are absent in birds :
(A) Teeth (B) Lip (C) Soft palate (D) All of the above

79. In which of the following bird crop is absent :
(A) Owl (B) Parrot (C) Pigeon (D) All of the above

80. How many lobes are present in the liver of birds :
(A) 2 (B) 3 (C) 4 (D) 5

81. How many lobes are present in the liver of birds :
(A) 2 (B) 3 (C) 4 (D) 5

82. Control of appetite in the birds and mammals is controlled by :
(A) Hypothalamus (B) Cerebral cortex
(C) Cerebellum (D) Medulla oblongata

83. Salivary gland secretion of poultry is which of the following type :
(A) Serous type (B) Mucus type
(C) Both (A) and (B) (D) None of these

84. Salivary gland secretion of pegion is which of the following type :
(A) Serous type (B) Mucus type
(C) Both (A) and (B) (D) None of these

85. Gastric pH of bird is :
(A) 0.5-2.5 (B) 3-5 (C) 5-6.5 (D) 6-8

86. Intestinal pH of birds is :
(A) 0.5-2.5 (B) 3-5 (C) 5-6.5 (D) 6-8

87. Pancreatic secretory rate of birds .... than animals.
(A) Higher (B) Lower
(C) Same (D) None of these

88. Absorption site of bile salt is which of the following :
(A) Lower ileum (B) Upper half of ileum
(C) Lower half ileum (D) All of these

89. Absorption site of endogenous amino acids is which of the following :

(A) Lower ileum (B) Upper half of ileum
(C) Lower half ileum (D) All of these

90. The stages of development of ruminant stomach in the embryo indicate that the part of the primitive spindle from which the abomasum develops :

(A) Dorsal (B) Caudal
(C) Proximal (D) None of the above

91. Bile is synthesized in :

(A) Liver (B) Gall bladder
(C) Pancreas (D) Duodenum

92. What is the role of gall bladder on bile :

(A) It dilutes the bile (B) It concentrates the bile
(C) It synthesizes bile (D) Both (A) and (B)

93. Color of bile in herbivorous is green and in carnivorous it is :

(A) Also green (B) Golden yellow
(C) Golden green (D) Golden black

94. Any substance that increases the secretion of bile is called :

(A) Choline (B) Chyme
(C) Cholerectic (D) All of the above

95. Up and down movement of gastric content before vomition is known as :

(A) Retching (B) Belching
(C) Projection (D) Eructation

96. Belching gas from rumen via mouth is known as :

(A) Retching (B) Belching
(C) Projection (D) Eructation

97. Belching is associated with :

(A) Secondary ruminal contraction (B) Extra ruminal contraction
(C) Both (A) and (B) (D) None of these

98. Regurgitation is associated with :

(A) Secondary ruminal contraction (B) Extra ruminal contraction
(C) Both (A) and (B) (D) None of these

99. Which of the following receptors responsible for eructation :

(A) Volume receptor of rumen (B) Stretch receptor of cardia
(C) Both (A) and (B) (D) None of these

100. The temperature of rumen is :

(A) 25-30°C (B) 36-38°C
(C) 40-42°C (D) 42-45°C

— Answer Sheet —

Q. No.	Ans	Q. No.	Ans	Q. No.	Ans	Q. No.	Ans	Q. No.	Ans	Q. No.	Ans
1.	**C**	18.	**C**	35.	**A**	52.	**B**	69.	**A**	86.	**D**
2.	**C**	19.	**D**	36.	**C**	53.	**A**	70.	**C**	87.	**A**
3.	**B**	20.	**B**	37.	**C**	54.	**B**	71.	**D**	88.	**A**
4.	**A**	21.	**A**	38.	**D**	55.	**D**	72.	**B**	89.	**C**
5.	**D**	22.	**B**	39.	**A**	56.	**A**	73.	**A**	90.	**B**
6.	**D**	23.	**C**	40.	**A**	57.	**B**	74.	**C**	91.	**A**
7.	**B**	24.	**B**	41.	**A**	58.	**C**	75.	**C**	92.	**B**
8.	**A**	25.	**A**	42.	**B**	59.	**D**	76.	**C**	93.	**B**
9.	**B**	26.	**D**	43.	**D**	60.	**D**	77.	**A**	94.	**C**
10.	**A**	27.	**D**	44.	**D**	61.	**A**	78.	**D**	95.	**A**
11.	**C**	28.	**D**	45.	**B**	62.	**C**	79.	**A**	96.	**D**
12.	**C**	29.	**A**	46.	**C**	63.	**A**	80.	**A**	97.	**A**
13.	**A**	30.	**D**	47.	**C**	64.	**A**	81.	**B**	98.	**B**
14.	**C**	31.	**B**	48.	**C**	65.	**C**	82.	**A**	99.	**C**
15.	**C**	32.	**C**	49.	**C**	66.	**D**	83.	**A**	100.	**B**
16.	**D**	33.	**A**	50.	**D**	67.	**B**	84.	**C**		
17.	**C**	34.	**B**	51.	**D**	68.	**D**	85.	**A**		

• Chapter 5 •

Endocrine System and Lactation Physiology

1. **The first hormone isolated in 1904 by Bayliss and Sterling is :**
   (A) Secretin (B) Epinephrine
   (C) Gastrin (D) Prolactin
2. **Secretin has ...... number of amino acids.**
   (A) 17 (B) 27 (C) 37 (D) 47
3. **Secretin is released into r intestinal lumen in response to :**
   (A) Lower duodenal PH (B) Higher duodenal PH
   (C) Presence of fat in duodenum (D) Presence of peptide in duodenum
4. **First hormone synthesized in lab is :**
   (A) Secretin (B) Insulin (C) Epinephrine (D) Gastrin
5. **The primary structure of bovine insulin was first determined by :**
   (A) Bayliss and Starling (B) F. Sanger
   (C) Leewenhock (D) Har Gobind Khorana
6. **Major form of thyroid hormone in blood is :**
   (A) Thyroxine (B) Thyroglobulin
   (C) Tri.iodothyronine (D) Both (A) and (C)
7. **T3 is .... times more active than T4.**
   (A) 5 times (B) 10 times (C) 15 times (D) 20 times
8. **For thyroid hormone receptors are located :**
   (A) On the cell membrane (B) In the cytosol
   (C) In the nucleus (D) All over the cell
9. **An autoimmune disease associated with hyperthyroidism :**
   (A) Grave's disease (B) Hashimoto's thyroiditis
   (C) Myxedema (D) Both (A) and (B)
10. **Goiter is associated with :**
    (A) Hypothyroidism (B) Hyperthyroidism
    (C) Iodine deficiency (D) All of the above
11. **Metabolic rates in mammals is controlled by :**
    (A) Liver (B) Pancreas
    (C) Thyroid (D) Pituitary
12. **Glucocorticoid is secreted by :**
    (A) Zona fasiculata (B) Zona reticularis
    (C) Zona glomerulosa (D) Adrenal medulla
13. **Which of the following organ secret Follicle secreting hormone (FSH) :**
    (A) Anterior pituitary (B) Ovary
    (C) Hypothalamus (D) Parathyroid
14. **In adults insufficient thyroxine will leads to :**
    (A) Cretinism (B) Acromegaly
    (C) Myxedema (D) Goiter

15. The endocrine gland which contributes to setting the body's biological clock is the :

(A) Pituitary (B) Hypothalamus
(C) Pineal (D) Parathyroid

16. Name the condition where the concentration of ketone bodies increases in urine :

(A) Diabetes insipidus (B) Diabetes mellitus
(C) Myxedema (D) Cushing syndrome

17. The enzyme required for synthesis of c-AMP is :

(A) ATPase (B) Adenylate cyclase
(C) Phospholipase C (D) G-protein

18. G-protein is discovered by Gliman and Rodbell while studying action of .... hormone.

(A) Secretin (B) Adrenalin (C) Insulin (D) Glucagon

19. GDP bound to .... subunit of G protein.

(A) α (B) β (C) γ (D) all the three

20. GH contains .... amino acids.

(A) 90 (B) 190 (C) 191 (D) 198

21. IGF-1 is also called :

(A) Somatostatin (B) Somatotropin
(C) Somatomedin (D) Somatoliberin

22. IGF-1 is secreted from :

(A) Hypothalamus (B) Pituitary (C) Liver (D) Kidney

23. Chromophil containing cells are present in :

(A) Zona glomerulosa (B) Zona fasciculate
(C) Zona reticularis (D) Adrenal medulla

24. Homeostasis was first recognized by :

(A) Bayliss and Sterling (B) Claude Bernard
(C) E.E. Muller (D) R.G.Smith

25. Which one is false about hormone function :

(A) The effects of hormone are short lasting and is not seen after levels of hormone return to basal levels
(B) Hormones are present in trace amount in plasma, usually ranging from 10^{-9} to 10^{-6} g/ml.
(C) They are sent in all times in order to maintain receptors in the target tissue.
(D) Hormones are secreted in variable amounts according to need.

26. Synergistic hormone action is seen is :

(A) Additive hormone action (B) Non.additive hormone action
(C) Permissive action (D) All of the above

27. Some hormones have no effect of their own but must be present for another hormone to have effect :

(A) Additive hormone action (B) Non.additive hormone action
(C) Permissive action (D) All of the above

28. .... is a best example of permissive hormone effect.
(A) Sex hormones and prolactin
(B) Thyroid hormone and growth hormone
(C) Steroid hormone and growth hormone
(D) all of these

29. When carrier proteins get bind to hormone what happens to its half life in circulation :
(A) Increased (B) Decreased
(C) Cannot be said (D) No change

30. Which among the following is not a glycoprotein :
(A) FSH (B) LH (C) TSH (D) MSH

31. Which among the following is false about receptor :
(A) Receptors are normally present in small numbers about 10,000 molecules per cell
(B) They are mainly 2 types i.e. cell-surface receptor and intra-cellular receptor
(C) Peptides and proteins enter in to cell to interact with intra-cellular receptor
(D) For some cell-surface receptors, a second messenger system is needed

32. .... is the major glucocorticoid in animal body.
(A) Cortisol (B) Aldosterone
(C) Corticosterone (D) Progestins

33. The synthesis of steroid hormones occurs at :
(A) Golgi apparatus
(B) Smooth Endoplasmic Reticulum
(C) Rough Endoplasmic Reticulum
(D) Ribosome

34. .... is the precursor of all steroid hormones.
(A) Acetate (B) Cholesterol
(C) Vegetable Oil (D) Mineral oil

35. Which among the following is not an eicosanoid hormone :
(A) Prostaglandin (B) Thromboxane
(C) Progestins (D) Prostacyclins

36. What are the eicosanoids present which have contradictory effect on blood clotting :
(A) PGF2α and PGI2 (B) PGI2 and TXA2
(C) Leukotrienes and TXA2 (D) None of these

37. Which among the following is false about peptidases :
(A) Exopeptidases degrade protein from carboxyl terminal end only
(B) Trypsin hydrolyses peptide bonds where the carboxyl group is from lysine or arginine
(C) Chymotrypsin hydrolyses peptide bonds where the carboxyl group is from phenylalanine, tryptophan or tyrosine
(D) Metabolism of proteins occurs mostly in liver, kidney and target cell lysosomes

38. Which one is false about steroid hormones :

(A) Physiologically only 5-10% steroid hormone is found in free or unbound form

(B) Steroids are bound by carrier proteins called albumin or steroid bounding globulin

(C) Steroids are degraded by a two-phase process in liver and then in kidney

(D) Steroids are water soluble

39. Which of the following describes clearly an insulin receptor :

(A) It has molecular mass of 200-400 kDa, consisting of two α subunits of 130 kDa and two β subunits of 90kDa

(B) It has molecular mass of 200-400 kDa, consisting of two α subunits of 130 kDa and two β subunits of 90kDa linked by disulphide bonds

(C) It has molecular mass of 200-400 kDa, consisting of two α subunits of 90kDa and two β subunits of 130 kDa

(D) It has molecular mass of 200-400 kDa, consisting of two α subunits of 90kDa and two β subunits of 130 kDa linked by disulphide bonds

40. Which among the following is a second messenger :

(A) ATP (B) cAMP

(C) Adenylate cyclase (D) Phospholipase C

41. Which among the following is false about calmodulin :

(A) It is a heat stable globular protein

(B) It has molecular weight of 16kDa

(C) It is a calcium dependent regulatory protein found in all eukaryotic cell

(D) One molecule of calmodulin can bind with 2 Ca^{2+} ions to form an active complex

42. Which among the following is false about tyrosine kinase receptor :

(A) This receptor do not use a second messenger system to activate a aseparate protein kinase

(B) It has a kinase domain which can activate the protein tyrosine kinase after hormone.receptor binding

(C) The hormone consist of a transmembrane and cytoplasmic domain but no extracellular domain

(D) The cytoplasmic domain has a C terminal tail with auto.phosphorylation site

43. Estrogen receptors were identified in 1966 by ....

(A) E.E. Muller (B) Toff and Gorski

(C) R.G.Smith (D) Bayliss and Sterling

44. Which among the following is false about intracellular receptors :

(A) Steroid, thyroid and most protein hormones have intracellular receptors

(B) Glucocorticoid, mineralocorticoid and androgens have predominantly cytoplasmic receptor and thyroid hormone, estrogen and progesterone have nuclear receptor

(C) Steroid hormone receptors in absence of hormone can move between nucleus and cytoplasm, are bound to HSP 90

(D) Binding of hormone to receptor releases HSP 90 and the Hormone-Receptor complex is translocated to nucleus

45. Which one is true about nuclear receptor ?
(A) They have 2 structural domains i.e. DNA binding and hormone binding domain
(B) To the hormone binding domain, hormone binds
(C) The DNA binding domain is comprised of 60-70 amino acids arranged in 'Zinc fingers'
(D) All of the above are true

46. Oxytocin and vasopressin are released from posterior pituitary, they have :
(A) Single amino acid (B) Two amino acids
(C) 9 amino acids (D) 39 amino acid

47. Which among the following is false about pituitary hormones ?
(A) Anterior pituitary produces number of releasing hormones
(B) Hormones of anterior pituitary are proteins or glycoproteins
(C) Neurons from hypothalamus directly innervated the posterior pituitary to release hormones
(D) Anterior pituitary hormones have longer half-life than hypothalamic releasing hormones

48. The anatomical joining of two individuals, especially artificially in physiological research is called :
(A) Parabiosis (B) Antibiosis (C) Symbiosis (D) Probiosis

49. Who among the following are inhibitors of microtubules and microfilaments ?
(A) Deoxyglucose and ionophores
(B) Colchicine and cytochalasin B
(C) β-galactosidase and neuraminidase
(D) Concavalin A and Forskolin

50. Valinomycin is an ionophore having specificity on .... transport.
(A) Ca^{2+} (B) Na^{+} (C) K^{+} (D) Cl^{-}

51. The unique site on the surface of antigen where antibody binds is called :
(A) Epitope (B) Paratope (C) Hepten (D) Liposome

52. A small molecule which by itself lacks antigenicity, but when combined with a larger carrier such as a protein, can elicit the production of antibodies which bind specifically to it :
(A) Epitope (B) Paratope (C) Hepten (D) Liposome

53. Assay system of Bioassay for somatotropin is :
(A) On frog skin
(B) On castrated or immature rodent
(C) Tibia test
(D) On rat uterus

54. 21-hydroxy pregn-4-ene-3,20-dione is the systemic name of :
(A) Aldosterone (B) 11-deoxy corticosterone
(C) Preguenolone (D) Progesterone

55. An animal that contains exogenous gene or gene modification is called :
(A) Transgenic animal (B) Hybrid animal
(C) Crossbred (D) Parthenogenic animal

56. Which one is false about broodiness behaviour ?

(A) Broodiness cannot be regulated by manipulation of environment as it is controlled purely by hormones and genotype

(B) An extensive brood patch develops on the breast for improved heat transfer from the hen to embryo

(C) Prolactin acts in concert with the ovarian steroids to increase vascularization and feather loss in brood patch

(D) Prolactin hormone is associated with broodiness

57. What is VIP ?

(A) Vascular inhibitory peptide (B) Vesicular inhibitory peptide

(C) Vasoactive intestinal peptide (D) Variable inhibitory peptide

58. The word pheromone was used by Karlson and Butenandt in 1959, it is derived from a Greek word "pherein" meaning :

(A) Smell (B) Transfer (C) Sex (D) Volatile

59. Any chemical substance produced and released by an individual of one species that affects the behaviour of a member of another species to the benefit of the originator but not the receiver is called :

(A) Kairomones (B) Spiromones (C) Allomones (D) Isomones

60. A chemical substance emitted by an organism and detected by another of a different species which gains advantage from this, e.g. a parasite seeking a host is called :

(A) Kairomones (B) Spiromones (C) Allomones (D) Isomones

61. Which among the following is true about Calcium and muscle contraction ?

(A) Calcium is required for muscle contraction as well as muscle relaxation

(B) Acetyl choline mediates contraction via opening the calcium channels of sarcoplasmic reticulum i.e. called Electro-mechanical coupling.

(C) The calcium ions then flow into the cytoplasm and bind to the troponin and tropomyosin molecules in the actin filaments of the muscle cells. The binding of calcium allows the troponin to change shape and expose the actin-myosin binding site.

(D) All are true

62. Pheromones are classified in to .... types.

(A) 2 (B) 3 (C) 4 (D) 5

63. Types of pheromones include :

(A) Controlling pheromone and driving pheromone

(B) Priming pheromone and signaling pheromone

(C) Physical pheromone and chemical pheromone

(D) All of the above

64. Which type of pheromone brings physiological changes that have long term influence ?

(A) Controlling Pheromone (B) Driving Pheromone

(C) Priming Pheromone (D) Signaling Pheromone

65. Dococenyl acetate is a sex attractant for :

(A) Sheep (B) Fox (C) Asian Elephant (D) Reindeer

66. .... is a family of protein act as a carrier for pheromones and play crucial role in its delivery and detection.

(A) Hepten (B) Lignocaine
(C) Lipocalins (D) Pregnenolone

67. The organ important for pheromone detection in mammals is :

(A) Sensilla (B) Hair cell
(C) Vomeronasal organ (D) All of the above

68. The pregnancy termination in rodent due to exposure of female to genetically different male :

(A) Bruce effect (B) Whitten effect
(C) Lee-Boot effect (D) Coolidge effect

69. Phenomenon seen in mammalian species whereby males (and to a lesser extent females) exhibit renewed sexual interest if introduced to new receptive sexual partners, even after cessation of sex with prior but still available sexual partners is called :

(A) Bruce effect (B) Whitten effect
(C) Lee-Boot effect (D) Coolidge effect

70. When male pheromones stimulate synchronous estrus in a female population ?

(A) Bruce effect (B) Whitten effect
(C) Lee-Boot effect (D) Coolidge effect

71. Phenomenon concerning the suppression or prolongation of oestrous cycles of mature female mice (and other rodents), when females are housed in groups and isolated from males :

(A) Bruce effect (B) Vandenberg effect
(C) Lee-Boot effect (D) Coolidge effect

72. Phenomenon in which an early induction of the first estrous cycle in prepubertal female mice occurs as a result of exposure to the pheromone-laden urine of a sexually mature (dominant) male mouse :

(A) Bruce effect (B) Vandenberg effect
(C) Lee-Boot effect (D) Coolidge effect

73. The boar pheromones cause the sow in estrus to stand for mating, this is called :

(A) Stability (B) Persistency (C) Lordosis (D) Boar mate

74. Mullerian duct differentiate in fetal life to form :

(A) Ovary (B) Oviduct (C) Uterus (D) All of the above

75. Puberty occurs at about .... month of age.

(A) 6-7 month (B) 12-14 month
(C) 16-18 month (D) 2 years of age

76. Who among the following are true seasonal breeder ?

(A) Sheep, goat, horse (B) Cattle, buffalo, sheep
(C) Dog, cow, horse (D) All of the above

77. .... has long acting FSH activity.

(A) hCG (B) PMSG
(C) LH (D) Pregnenolone

78. Which among the following is a synthetic progesterone ?
(A) Melengestrol acetate
(B) 6-methyl-17-acetoxy-progesterone
(C) 6-chloro-8-dehydro-17-acetoxy-progesterone
(D) All of the above

79. Seasonal singing in birds is mainly influenced by :
(A) Testosterone (B) Prolactin
(C) Progesterone (D) Oxytocin

80. Initiation of lactation is also called as :
(A) Mammogenesis (B) Lactogenesis
(C) Galactopoeisis (D) Let down

81. When certain body parts grows at a faster rate compared to other body part at a particular physiological stage for ex. After puberty mammary gland grows at a faster rate than other body parts it is called ?
(A) Heterometric growth (B) Allometric growth
(C) Unsynchronized growth (D) Asymmetric growth

82. Mammary duct development increases by around .....% in each estrous cycle :
(A) 2 (B) 5 (C) 8 (D) 16

83. The lactating mammary gland of dairy cattle weighs :
(A) 2-5 kg (B) 15-25 kg
(C) 25-35 kg (D) 35-50 kg

84. The lobulo.alveolar growth of mammary gland is stimulated by :
(A) Estrogen only (B) Progesterone only
(C) Prolactin only (D) Both estrogen and progesterone

85. Lactogenesis is initiate by the complex action of :
(A) Estrogen, progesterone and somatotropin
(B) Prolactin, estrogen and progesterone
(C) Prolactin, estrogen and glucocorticoid
(D) Prolactin, progesterone and oxytocin

86. Which among the following is true about galactopoeisis ?
(A) Thyroid hormones are not required for maximal milk production
(B) High levels of exogenous glucocorticoids and estrogen decrease milk production in an established lactation
(C) Prolactin is required for the maintainace of lactation in rats and rabbit not in cows
(D) Placental lactogen administration in lactating cows increases their milk yield

87. IGF-1 is synthesized in :
(A) Pituitary (B) Hypothalamus
(C) Liver (D) All of the above

88. The main compound involve in development of boar taint is :
(A) Androstenone (B) Skatole
(C) Both the above (D) Steroids

89. Which statement is true about thyroid hormone ?

(A) Thyroid hormone is synthesized in the follicles of thyroid gland

(B) Thyroid hormones are highly hydrophobic and are bound to transthyretin and other proteins in blood for transport

(C) T4 is converted to biologically most active form i.e. T3 outside the thyroid gland by the help of Type I and Type II 5' deiodinase

(D) All of the above are true

90. Which statement is false about leptin hormone ?

(A) Leptin is a 16kDa protein hormone produced by brown adipose tissue.

(B) It is a product of "ob" obese gene

(C) It act like an 'adipostat' and important regulator of appetite.

(D) Plasma leptin varies in a pulsatile manner, with approx. 30 pulses in 24 hours in human.

91. Which among the following is false about cholecystokinin (CCK) ?

(A) CCK is a peptide hormone produced by duodenal mucosa

(B) CCK stimulates pancreatic enzyme secretion and bile secretion

(C) CCK is structurally similar to gastrin enzyme

(D) CCK increases appetite

92. Which among the following is false about Insulin ?

(A) It is a polypeptide hormone produced by β-cells in the islets of Langerhans in pancreas

(B) Its primary role is to regulate blood glucose level by increasing uptake of glucose in to tissue and store them in form of glycogen and lipid

(C) Chromium inproves the function of insulin

(D) Insulin consist of two polypeptide chains linked to each other by a carbohydrate chain

93. Which among the following is false about Glucagon ?

(A) t is a peptide hormone produced by α-cells in the islets of Langerhans in pancreas

(B) It has two polypeptide chain of 29 amino acid each linked by disulphide bridge.

(C) The effects of glucagon is opposite to insulin

(D) This hormone function is important during prolong fasting, exercise and during neonatal life.

94. What is xenobiotic ?

(A) Analog of antibiotic

(B) Similar to probiotic

(C) Foreign chemicals found in human body which are not naturally found

(D) Both probiotic and prebiotic collectively called xenobiotic

95. Which one is false about T cells ?

(A) There are four groups of T lymphocytes

(B) T helper cells produce lymphokines

(C) Cytotoxic T killer cells attack and kill foreign cells

(D) Suppressor T cells regulate other cells to prevent severe immune reaction that could damage animal's own tissue

96. What is allostasis ?
(A) It is the adaptive response to stress
(B) It is a successful skin grafting phenomena from other individual
(C) It is the accumulation of food other than cholesterol in blood vessels
(D) None of the above is true

97. Progesterone is produced by :
(A) Ovary (B) Placenta
(C) Adrenal gland (D) All of the above

98. Which among the following is not a short-day breeder ?
(A) Sheep (B) Goat (C) Horse (D) Deer

99. With maturation of Hypothalamic-Pituitary axis :
(A) Both the amplitude and frequency of LH increases
(B) Both the amplitude and frequency of LH decreases
(C) Both the amplitude and frequency of estrogen increases
(D) Both the amplitude and frequency of estrogen decreases

100. Which among the following is false about puberty ?
(A) Onset of puberty is influenced by target age, weight, breed, nutritional status and environment
(B) Puberty occurs at around 6-7 months of age in sheep, goat and pigs and at around 12 months age in cattle
(C) During the breeding season the negative feedback of estradiol on LH is reduced
(D) Seasonal differences in photoperiod is recognized by adrenal gland

101. Which among the following is false about melatonin ?
(A) The pineal gland releases melatonin
(B) It is secreted during the period of darkness
(C) The pattern of melatonin release regulates seasonal rhythms in only short day breeders but not in long day breeders
(D) Melatonin does not act directly on the GnRH neurons, but act indirectly on other neurons that synapse with GnRH neurons

102. Important role of somatostatin :
(A) Stimulates protein deposition in muscles
(B) Stimulates lipogenesis in adipose tissue
(C) Reduces protein deposition in muscles
(D) Stimulates feed intake

103. Regarding thyroid hormone synthesis which one is correct :
(A) Predominant hormone made by thyroid follicular cells is T 3
(B) Thyroid hormone is actively transported from thyroid follicular epithelial cells in to the circulation by a specific transport protein
(C) Iodine that is released from proteolytic cleavage of thyroglobulin can be recycled from mono and di iodotyrosines
(D) The synthesis of T4 occurs within the thyroid follicular cells

104. Which adrenal cortical hormone is absent in poultry ?
(A) Aldosterone (B) Cortisol
(C) Corticosterone (D) Dihydrocorticosterone

105. 1, 25 dihydroxy cholecalciferol formed in :
(A) Liver (B) Kidney (C) Pancreas (D) None

106. Homeorhesis is coined by :
(A) Bauman and Currie (B) Claude Bernard
(C) Toff and Gorski (D) None of the above

107. Mammary gland development in early embryonic life starts from :
(A) Ectoderm (B) Endoderm
(C) Mesoderm (D) Yolk

108. During the embryonic life first structure that develop from ectoderm :
(A) Mammary band (B) Mammary bud
(C) Mammary hillock (D) Mammary streak

109. In how many days mammary hillock develops in embryo :
(A) 7 (B) 10 (C) 20 (D) 40

110. In how many days mammary band develops in embryo :
(A) 32 (B) 34 (C) 36 (D) 37

111. For normal milk synthesis which structural unit is essential :
(A) Fat pad (B) Fibrous connective tissue
(C) Myoepithelial cell (D) Alveolar cells

112. At what structural developmental stage mammary differentiation ceases in male :
(A) Mammary band (B) Mammary bud
(C) Primary sprout (D) Secondary sprout

113. From which embryonic structure gland cistern and teat cistern develops :
(A) Mammary band (B) Mammary bud
(C) Primary sprout (D) Secondary sprout

114. Connective tissue of mammary gland develops from :
(A) Ectoderm (B) Mesoderm
(C) Endoderm (D) Both (A) and (B)

115. Gland and teat cistern develops at .... age of embryonic life.
(A) 34 and 37 days (B) 50 and 55 days
(C) 78 and 90 days (D) 110 and 130 days

116. Prolactin is luteotropic in which species :
(A) Mice (B) Rat
(C) Hamster (D) All of the above

117. Mammary duct lengthening, branching and thickening occurs due to :
(A) Estrogen alone
(B) Progesterone alone
(C) Estrogen and progesterone
(D) Estrogen, progesterone, prolactin and somatotropin

118. Apart from four normal teats some cows have an additional teat called as :
(A) Additional teat (B) Rear teat
(C) Supernumerary teat (D) Pseudo teat

119. Which suspensory structure is important in absorption of shock when the cow walks ?
(A) Medial suspensory ligament (B) Lateral suspensory ligament
(C) Prepubic tendon (D) All of the above

120. The only opening of the teat that connects the secretary system of mammary gland with external environment :
(A) Streak canal (B) Teat cistern
(C) Furstenburg's Rosette (D) Both (A) and (C)

121. For 1 liter milk production how much blood is needed :
(A) 3-4 liter (B) 40-50 liter
(C) 400-500 liter (D) 3000-4000 liter

122. Why buffaloes are called hard milker ?
(A) Takes more time for milk let down
(B) Difficult to handle due to large body size
(C) Gives less milk
(D) All are true

123. Why milk let down in difficult in buffaloes ?
(A) Most milk (90-95%) is cisternal
(B) Sphincter muscle closure at teat opening is more stronger
(C) Both (A) and (B)
(D) Less oxytocin in circulation

124. Interlobar ducts open into a large basin called as :
(A) Gland cistern (B) Teat cistern
(C) Lactiferous sinus (D) Both (A) and (B)

125. Smallest duct in mammary tissue is called
(A) Interlobar duct (B) Interlobular duct
(C) Intercalary duct (D) Both (B) and (C)

126. Arteries supplying the teat is called :
(A) Caudal mammary artery (B) External pudendal artery
(C) Prepubic artery (D) Papillary artery

127. Which among the following is known as milk vein ?
(A) Subcutaneous abdominal vein (B) External pudic vein
(C) Femoral vein (D) Common iliac vein

128. During allometric growth of mammary tissue mammary cells grow at a .... pace/rate higher than other body tissue.
(A) 1-5 times (B) 2-5 times
(C) 3-5 times (D) 4-5 times

129. During every estrous cycle approximately .... percent ductal development occur.
(A) 2 (B) 4 (C) 6 (D) 8

130. Placental lactogen (PL) has properties and chemical makeup similar to :
(A) Prolactin (PRL) (B) Growth Hormone
(C) Relaxin (D) Both (A) and (B)

131. Lobulo-alveolar growth of mammary gland is due to :

(A) Estrogen (B) Progesterone
(C) Somatotropin (D) Placental lactogen

132. Which among the following animals do not have placental lactogen ?

(A) Pig and goat (B) Goat and rabbit
(C) Rabbit and pig (D) Goat and horse

133. In rabbit important mammogenic hormone during pregnancy is :

(A) Growth hormone (B) Estrogen
(C) Prolactin (D) Relaxin

134. In sow important mammogenic hormone during pregnancy is :

(A) Growth hormone (B) Estrogen
(C) Prolactin (D) Relaxin

135. During first few days of lactation .... is the important signal for increase mitosis of mammary secretory cells.

(A) Suckling
(B) Suckling and milk removal
(C) Suckling, milk removal and playing music
(D) No mitosis occur after calving

136. Milking machine creates :

(A) Negative pressure (B) Positive pressure
(C) Both (D) Varies

137. Which among the following nucleus have cellbody containing oxytocin ?

(A) Suproptic and paraventricular nucleus
(B) Preoptic and supraoptic nucleus
(C) Preoptic and paraventricular nucleus
(D) Preoptic, supreoptic and paraventricular nucleus

138. Oxytocin during milk letdown act on :

(A) Alveolar cell (B) Fatty cell
(C) Myoepithelial cell (D) All of the above

139. Oxytocin takes .... time to cause milk letdown.

(A) 5-10 second (B) 20-40 second
(C) 40 second to 2 minutes (D) 2 minutes to 5 minutes

140. Oxytocin is stored in posterior pituitary in a bound form with a carrier protein called :

(A) Oxypoeitin (B) Neurophysin
(C) Oxyphysin (D) Neuropoeitin

141. Oxytocin.neurophysin bound molecule slowly trickle down from hypothalamus to posterior pituitary at a speed of :

(A) 2-3 mm/h (B) 5-10 mm/h
(C) 2-3 cm/h (D) 5-10 cm/h

142. Peripheral inhibition is caused by :

(A) Cortisol (B) Catecholamines
(C) Endorphins (D) All of these

143. Oxytocin and ADH differ in .... amino acid composition.
(A) 1 (B) 2 (C) 3 (D) 4

144. Which among the following is the role of oxytocin :
(A) Milk letdown
(B) Milk letdown and Maternal behaviour
(C) Milk letdown, Maternal behaviour and intense contraction during parturition
(D) Milk letdown, Maternal behaviour, intense contraction during parturition and osmoregulation

145. Death of calf of a cows results :
(A) Central inhibition (B) Peripheral inhibition
(C) No inhibition (D) Both (A) and (B)

146. What is the full form of FIL :
(A) Feedback initiation of lactation (B) Feedback inhibitior of lactation
(C) Full initiation of lactation (D) Full inhibition of lactation

147. Which among the following s false about FIL :
(A) It is a milk borne glycoprotein synthesized by secretory cells
(B) It has no inhibitory effect on the cell it is produced
(C) It is only active within alveoli and its effect depends upon the concentration
(D) FIL is present in more concentration in the alveoli where milk does not come out completely

148. Oxytocin is used in treatment of :
(A) Retention of placenta
(B) Retention of placenta and uterine involution
(C) Retention of placenta, uterine involution and Mastitis
(D) None of these

149. The amount of milk left in udder even after a normal milking and obtained only after the oxytocin injection is called :
(A) Leftover milk (B) Residual milk
(C) Cisternal milk (D) All of the above

150. Fat percent of Jersey cow is approximately :
(A) 3% (B) 3-5% (C) 4% (D) 5%

151. Colostrum contain............ in less quantity that milk.
(A) Fat (B) Protein (C) Lactose (D) Sodium

152. Dry period is given for
(A) 30 days (B) 45 days (C) 60 days (D) 90 days

153. Residual milk accounts .... % of milk in udder.
(A) 1-5% (B) 5-10% (C) 10-20% (D) 25-40%

154. Which among the following is true about residual milk ?
(A) The amount of residual milk is directly proportional to amount of milk present in the udder at the beginning of milking
(B) Older cows have higher percentage of residual milk than young cows
(C) The percentage of residual milk is higher in low producer than higher producer
(D) All of the above are true

155. The secretion rate stops completely :
(A) Without milking the cow for 35 hours
(B) After development of 60-70 mm Hg udder pressure
(C) Both the above are correct
(D) Without milking the animal for 24 hours or with development of 45mm Hg udder pressure

156. Anatomical defense mechanism in udder include :
(A) Sphincter muscle
(B) Sphincter muscle and Furstenberg's Rosette
(C) Sphincter muscle, Furstenberg's Rosette and keratin lining of streak canal
(D) Sphincter muscle, Furstenberg's Rosette, keratin lining of streak canal and Gland cistern

157. Chemical defense of mammary gland include :
(A) Lactoferrin
(B) Lactoferrin and Lysosome
(C) Lactoferrin, Lysosome and Lactoperoxidase-thiocyanate-H_2O_2
(D) None of these

158. During lactogenesis :
(A) Estrogen secretion increases and progesterone decreases
(B) Estrogen and prolactin secretion increases and progesterone decreases
(C) Estrogen, Prolactin and Cortisol secretion increases and progesterone secretion decreases
(D) Estrogen, Prolactin, Cortisol and Epinephrine secretion increases and progesterone secretion decreases

159. Lactogenesis occurs in how many phases :
(A) Single (B) Two (C) Three (D) Five

160. Which one is incorrectly matched (hormone and their role in lactogenesis) ?
(A) Cortisol – differentiation of Rough endoplasmic reticulum and Golgi apparatus
(B) PRL – Increases the metabolism of mammary epithelial cells
(C) Progesterone – reduces the synergism between estrogen and PRL
(D) Estrogen – inhibit the secretion of PRL

161. What are the autocrine factor involved in galactopoiesis ?
(A) Removal of milk at regular interval (B) Frequency of milking
(C) Time of pregnancy (D) Both (A) and (B)

162. Process by which mammary gland returns to non.lactating stage is called :
(A) Induction (B) Involution
(C) Ejection (D) Drying

163. In non.ruminants biosynthesis of milk fat occurs in :
(A) Mitrochondria (B) Ribosome
(C) Endoplasmic Reticulum (D) Cytosol

164. In ruminants biosynthesis of milk fat occurs at :
(A) Mitrochondria (B) Cytosol
(C) Endoplasmic Reticulum (D) Both (B) and (C)

165. What is the normal pH of bovine milk ?
(A) 6 (B) 6.7 (C) 7 (D) 7.4

166. Milk proteins are often subdivided in to two classes i.e. casein and whey protein i.e. the insoluble and soluble fraction at pH :
(A) 4-6 (B) 5 (C) 5-3 (D) 6-7

167. Out of the following minerals which one has lower concentration in colostrum than milk :
(A) Sodium (B) Potassium (C) Calcium (D) Chloride

168. Which protein plays essential role in lactose biosynthesis ?
(A) Alpha s 1 Casein (B) Beta protein
(C) Alpha –Lactalbumin (D) Beta– Lactalbumin

169. Which animal milk lacks or contains very less lactose ?
(A) Horse (B) Kangaroo (C) Elephant (D) Seal

170. Which among the following is iron binding protein ?
(A) Lactoferrin and Transferrin (B) Transferrin and Apoferrin
(C) Lactoferrin and Apoferritin (D) All of the above

171. In fat globule membrane .... enzyme and .... protein is present.
(A) Xanthine oxidase and Butyrophyllin
(B) Xanthine oxidase and Casein
(C) Peroxidase and Butyrophyllin
(D) Peroxidase and Casein

172. Which among the following is true about fat globule membrane ?
(A) It protects fat molecules from flocculation and coalescence
(B) This membrane is derived from apical membrane of secretory cells probably from Golgi apparatus
(C) In general this membrane consist of 60% protein and nearly 40% phospholipids
(D) All of the above

173. Which step does not belongs to Involution ?
(A) Active involution
(B) Passive involution
(C) Steady stage involution
(D) Redevelopment and colostridogenesis

174. Fatty acid synthtase complex contains how many different enzymes :
(A) 2 (B) 4 (C) 7 (D) 10

175. The fatty acids found in different mammalian tissue is predominantly palmitic acid whereas in mammary tissue it is of short and medium chain fatty acid, which is probably due to presence of .... enzyme in mammary gland.
(A) Desaturase (B) Thioesterase I
(C) Thioesterase II (D) All of the above

176. Osmotic pressure regulator of milk include :
(A) Lactose (B) Sodium + Potassium
(C) Chloride (D) All of the above

177. Lactogenic complex include :
(A) Insulin, glucocorticoid and Prolactin
(B) Growth hormone, Insulin and Cortisol
(C) Prolactin, Cortisol and Relaxin (D) All of the above

178. Secretion where migration of secretory products through apex of epithelial cell without damaging the membrane is called :
(A) Apocrine (B) Merocrine (C) Eccrine (D) Holocrine

179. Lactose synthesis occurs at :
(A) Endoplasmic reticulum (B) Golgi body
(C) Cytosol (D) Nucleolus

180. .... is a negative regulator of alpha Lactalbumin.
(A) Estrogen (B) Progesterone
(C) Prolactin (D) Placental Lactogen

181. Lactose synthase enzyme includes :
(A) Alpha Lactalbumin and Galactosyltransferase
(B) Alpha Lactalbumin and Beta lactalbumin
(C) Alpha Lactalbumin, Beta lactalbumin and Galactosyltransferase
(D) Alpha Lactalbumin, Beta lactalbumin, Galactosyltransferase and UDP-Galactose

182. The ration of Acetyl Co A to Malonyl Co A during fatty acid synthesis determines :
(A) Quantity of fat in milk (B) Length of fatty acid formed
(C) Saturation of fatty acid formed (D) All of the above

183. In ruminants glucose sparing effect occur during fatty acid synthesis in mammary tissue due to absence of :
(A) Acetyl Co A Synthase Enzyme (B) Acyl Co A Synthase Enzyme
(C) Citrate Lyase Enzyme (D) Thioesterase II Enzyme

184. .... is the major galactopoietic hormone in non.ruminant.
(A) Estrogen (B) Placental Lactogen
(C) Prolactin (D) Glucocorticoid

185. Which among the following is true about Lactoperoxidase-thiocyanate-H_2O_2 system ?
(A) It is inhibitory to gram positive bacteria and is bactericidal to gram negative bacteria
(B) It modifies the sulphydry l group of bacterial cell membrane, which is otherwise important for glucose transport in bacteria
(C) Hypothiocyanate is produced
(D) All the above are true

186. Hormone responsible for holding up of milk is :
(A) Vasopressin (B) Oxytocin
(C) Adrenaline (D) All of the above

187. Maintenance of lactation is known as :
(A) Mammogenesis (B) Lactogenesis
(C) Galactogenesis (D) Galactopoeisis

188. The bacterial content will be maximum in :
(A) Whole milk (B) Fore milk (C) Stripping (D) Middle milk

189. The pH of freshly drawn milk is :
(A) 6 (B) 6.6 (C) 7 (D) 7.4

190. Colostrum is important for the newborn calves because it contain :
(A) IgA (B) IgD (C) IgE (D) IgM

191. Strippings are rich in :
(A) Fat (B) Protein (C) Vitamins (D) Lactose

— Answer sheet —

Q. No.	Ans	Q. No.	Ans	Q. No.	Ans	Q. No.	Ans	Q. No.	Ans	Q. No.	Ans
1.	A	33.	B	65.	C	97.	D	129.	D	161.	D
2.	B	34.	B	66.	C	98.	C	130.	D	162.	B
3.	A	35.	C	67.	C	99.	A	131.	A	163.	A
4.	B	36.	B	68.	A	100.	D	132.	C	164.	D
5.	B	37.	A	69.	D	101.	C	133.	C	165.	B
6.	A	38.	D	70.	B	102.	A	134.	D	166.	A
7.	A	39.	B	71.	C	103.	C	135.	B	167.	B
8.	C	40.	B	72.	B	104.	B	136.	A	168.	C
9.	A	41.	D	73.	C	105.	B	137.	A	169.	D
10.	D	42.	C	74.	B	106.	A	138.	C	170.	A
11.	C	43.	B	75.	A	107.	A	139.	C	171.	A
12.	A	44.	A	76.	A	108.	A	140.	B	172.	D
13.	A	45.	D	77.	B	109.	D	141.	A	173.	B
14.	C	46.	C	78.	D	110.	A	142.	B	174.	C
15.	C	47.	A	79.	A	111.	D	143.	B	175.	C
16.	B	48.	A	80.	B	112.	C	144.	D	176.	D
17.	B	49.	B	81.	B	113.	C	145.	A	177.	A
18.	B	50.	C	82.	C	114.	B	146.	B	178.	B
19.	A	51.	A	83.	B	115.	D	147.	B	179.	B
20.	C	52.	C	84.	D	116.	D	148.	C	180.	B
21.	C	53.	C	85.	C	117.	D	149.	B	181.	A
22.	C	54.	B	86.	A	118.	C	150.	D	182.	B
23.	D	55.	A	87.	C	119.	A	151.	C	183.	C
24.	B	56.	A	88.	C	120.	A	152.	C	184.	C
25.	A	57.	C	89.	D	121.	C	153.	C	185.	D
26.	B	58.	B	90.	A	122.	A	154.	D	186.	C
27.	C	59.	C	91.	D	123.	C	155.	C	187.	D
28.	C	60.	A	92.	D	124.	C	156.	C	188.	B
29.	A	61.	D	93.	B	125.	C	157.	C	189.	B
30.	D	62.	A	94.	C	126.	D	158.	C	190.	A
31.	C	63.	B	95.	A	127.	A	159.	B	191.	A
32.	A	64.	C	96.	A	128.	C	160.	D		

• Chapter 6 •
Respiratory and Excretory System

1. **Boyle's law explains :**
   (A) V α T (B) V α 1/P
   (C) V α Solubility (D) None of these
2. **Which among the following is true about solubility ?**
   (A) CO_2 is more soluble than oxygen
   (B) N_2 is more soluble than oxygen
   (C) Oxygen is more soluble than CO_2
   (D) Both (A) and (B) are true
3. **CO_2 is .... times more soluble than oxygen.**
   (A) 4 times (B) 9 times (C) 12 times (D) 22 times
4. **Fick's law explains about :**
   (A) Relationship of volume of gas with pressure
   (B) Factors linked with partial pressure of gases
   (C) Factors affecting gas diffusion between 2 surface
   (D) All of the above
5. **Partial pressure of oxygen in arterial blood is :**
   (A) 40 mm Hg (B) 45 mm Hg (C) 100 mm Hg (D) 149 mm Hg
6. **Muscle helps in expiration are**
   (A) External intercostal muscle
   (B) Internal intercostal and abdominal muscle
   (C) Diaphragm (D) Both (A) and (C)
7. **Inspiratory cycle is smooth and continuous, except in :**
   (A) Horse (B) Buffalo (C) Pig (D) Goat
8. **The compensatory mechanism associated with poor ventilation is called :**
   (A) Sigh (B) Compensatory breathing cycle
   (C) Costal breathing (D) Both (A) and (B)
9. **Normal quiet respiration is called :**
   (A) Abdominal breathing (B) Costal breathing
   (C) Sigh (D) None of these
10. **Which statement is true about intrathoracic pressure :**
    (A) Also called intrapulmonic pressure
    (B) Air pressure in lungs or passage leading to it
    (C) Always less than intra.alveolar pressure
    (D) All are true
11. **Effect of surface tension on pulmonary alveoli can be explained by :**
    (A) Fick's law (B) Laplace's law
    (C) Henry's law (D) Charle's law
12. **Lipid portion of pulmonary surfactant is :**
    (A) Dipalmitoyl choline (B) Dipalmitoyl lecithin
    (C) Stearic acid (D) Phosphotidyl choline

13. Pulmonary surfactants were synthesized by :

(A) Liver cell (B) Type I alveolar cells

(C) Type II alveolar cells (D) Type III alveolar cells

14. In horse and swine deficit of pulmonary surfactants result in :

(A) Sigh (B) Barker's syndrome

(C) Emphysema (D) Both and b

15. Pressure volume curve showing the compliance of lungs, shows lung is more compliant at :

(A) Beginning of pressure (B) Middle of pressure

(C) Towards the end of pressure (D) Equal at all the above conditions

16. Amount of air breathed in or out in each respiration is called :

(A) Vital capacity (B) volume

(C) Inspiratory volume (D) Tidal volume

17. Vital capacity is obtained by :

(A) Inspiratory capacity + tidal volume

(B) Inspiratory capacity + expiratory reserve volume

(C) Inspiratory capacity + minute volume

(D) Vital capacity + residual volume

18. What is true about dead space ventilation ?

(A) Anatomical dead space volume is fixed

(B) Air in the terminal bronchi and alveoli but unable to exchange with blood constitutes physiological dead space

(C) Dead space ventilation is important during thermoregulation

(D) All are true

19. Respiratory alkalosis results due to :

(A) Hyperventilation (B) Hypoventilation

(C) Normoventilation (D) Any of these

20. Hypoxic vasoconstriction is primarily due to :

(A) Decrease in PO_2 in alveolar gas

(B) Decrease in PO_2 in arterial gas

(C) Decrease in PCO_2 in alveolar gas

(D) Decrease in PCO_2 in arterial gas

21. Major amount of CO_2 is transported in :

(A) Dissolved form (B) Bicarbonate ion form

(C) Carbamino compound form (D) Both (A) and (B)

22. Tidal volume is measured by :

(A) Spectrometry (B) Spirometry

(C) Spectrophotometry (D) Anemometry

23. Which among the following is not a vasoconstrictor ?

(A) Thromboxane A2 (B) Endotheline

(C) Angiotensin II (D) Prostacyclin

24. Collateral ventilation is a prominent feature in :

(A) Dog (B) Horse (C) Pig (D) Cattle

25. One mole of haemoglobin can combines with :
(A) 2 moles of oxygen (B) 4 moles of oxygen
(C) 6 moles of oxygen (D) 8 moles of oxygen

26. Volume of CO_2 carried by blood in different form at different partial pressure of CO_2 is given by :
(A) Oxygen haemoglobin dissociation curve
(B) CO_2 dissociation curve
(C) Hamberger's phenomena
(D) Bohr's effect

27. The respiratory control system is present at :
(A) Hypothalamus (B) Median eminence
(C) Cerebellum (D) Brain stem

28. Inspiratory activity or basic rhythm of respiration is mainly control by :
(A) Dorsal respiratory group (B) Ventral respiratory group
(C) Rostral pons (D) Apneustic center

29. Output from dorsal respiratory group (DRG) to diaphragm is via :
(A) Vagus nerve (B) Glossopharyngeal nerve
(C) Phrenic nerve (D) Both (A) and (B)

30. Input to dorsal respiratory group (DRG) is via :
(A) Vagus nerve (B) Glossopharyngeal nerve
(C) Phrenic nerve (D) Both (A) and (B)

31. Neurons of ventral respiratory groups are primarily associated with :
(A) Inspiration (B) Expiration
(C) Deep inspiration (D) Respiration rate

32. Neurons of apneustic center are primarily associated with :
(A) Inspiration (B) Expiration
(C) Deep inspiration (D) Respiration rate

33. The receptors for Hering-Breuer reflexes are found in :
(A) Trachea (B) Bronchi and bronchioles
(C) Alveoli (D) Pulmonary blood vessels

34. The "J" receptors are found in :
(A) Trachea (B) Bronchi and bronchioles
(C) Alveoli (D) Pulmonary blood vessels

35. Which among the following is true about neural control of respiration ?
(A) Juxtacapillary receptors are present in alveoli and they are associated with increase in respiratory frequency
(B) Load receptors are present in inspiratory muscles like diaphagram
(C) Tendons and joints also have receptors associated with regulation of respiration
(D) All are true

36. Reduction in CO_2 and H^+ ion results decrease in ventilation is known as :
(A) Bohr's effect (B) Haldane's effect
(C) Braking effect (D) Hamberger's effect

37. Chloride shift is also known as :

(A) Bohr's effect (B) Haldane's effect
(C) Braking effect (D) Hamberger's effect

38. About .... percent of CO_2 in blood is transported in form of bicarbonate.

(A) 20 (B) 50 (C) 70 (D) 90

39. Bronchial circulation is ..... percent of cardiac output.

(A) 2 (B) 8 (C) 20 (D) 25

40. Which one is true about avian respiration ?

(A) Is circular instead of "C" shape as in mammals
(B) Tidal volume (VT) is 1.7 times less than mammals
(C) Like mammals bronchi branches several times to give the bronchial tree
(D) All are true

41. Which one is true about parabronchi ?

(A) Parabronchi are tertiary bronchi
(B) There are 2 different types of parabronchi
(C) Paleopulmonic parabronchi is present in all birds
(D) All are true

42. Which one is true about avian respiration ?

(A) Birds lack diaphragm
(B) Inspiration ad expiration are active process
(C) During inspiration pressure within the air sac decreases
(D) All are true

43. Surfectants in birds produced by :

(A) Type I alveolar cells
(B) Type II alveolar cells
(C) Cells of Ostia
(D) Trilaminar cells

44. Which one is false about avian haemoglobin (Hb) ?

(A) Avian haemoglobin is 2 types Hb A and Hb D
(B) Hb A has lower oxygen affinity
(C) Hb D is more predominant
(D) Hill's coefficient is 4

45. Purring is a respiratory phenomenon seen in :

(A) Dog (B) Cat (C) Horse (D) Pig

46. Increase frequency and depth of respiration is called as :

(A) Hypercapnia (B) Hyperpnea (C) Tachypnea (D) Polypnea

47. Hypoxia along with hypercapnia is called as :

(A) Atelectasis (B) Pneumonia (C) Asphyxia (D) Tachypnea

48. Methaemoglobinemia is an example of :

(A) Ambient hypoxia (B) Anemic hypoxia
(C) Stagnant hypoxia (D) Histotoxic hypoxia

49. At high altitude urine becomes :

(A) Acidic (B) Alkaline (C) Neutral (D) Unaltered

50. Which one is true about caisson's disease ?

(A) Seen in sea divers/ in water
(B) Air/ gas emboli are formed
(C) Also called decompression sickness
(D) All are true

51. What is true about hypobaric physiology ?

(A) Experienced by mountaineers (B) Associated with Boyle's law
(C) Associated with Fick's law (D) Both (A) and (C)

52. Inhalation/inspiration occurs as a result of :

(A) an upward movement of the diaphragm
(B) movement of the ribs closer together due to contraction of the inspiratory/inhalatory intercostal muscles
(C) a downward movement of the diaphragm
(D) Both (B) and (C)

53. In order for the lungs to function normally, the intrapleural pressure must :

(A) be lower than alveolar pressure
(B) be between +5 and +10 mmHg above atmospheric pressure
(C) be the same as atmospheric pressure
(D) Change as the respiratory demands of the body change

54. Oxygen is carried in blood :

(A) bound to haemoglobin
(B) dissolved in the cytosol of erythrocytes
(C) dissolved in the plasma
(D) in all the above forms

55. Which of the following would cause a decrease in the binding affinity of haemoglobin for oxygen ?

(A) Increased pH of the blood
(B) Increased temperature of the blood
(C) Decreased DPG levels in erythrocytes
(D) Both (A) and (C)

56. Regarding the airways and alveoli :

(A) The trachea, bronchi and bronchioles are all prevented from collapsing by the cartilage in their walls
(B) The small diameter airways are the main sites of airways resistance
(C) The upper airways play an important role in protecting the lungs from airborne particles
(D) The alveoli are the only site of gas exchange

57. Concerning the chemical regulation of respiration :

(A) The depth and rate of ventilation is decreased when an animal breathes air containing pure oxygen at normal atmospheric pressure.
(B) The central chemoreceptors sense the oxygen tension of the arterial blood
(C) The peripheral chemoreceptors are located in the carotid sinus and aortic arch
(D) The peripheral chemoreceptors are the only receptors able to respond to changes in the partial pressure of oxygen in the arterial blood.

58. Concerning ventilation of the lungs :

(A) The anatomical dead space is the volume of air taken in during a breath that does not enter the alveoli.

(B) The physiological dead space is always greater than the anatomical dead space

(C) The mean pressure in the pulmonary arteries rises as cardiac output increases

(D) The pattern of pulmonary blood flow is independent of posture

59. Regarding the pulmonary circulation :

(A) The pressures in the pulmonary arteries are similar to those in the systemic arteries

(B) The resistance of the pulmonary circulation rises as the pulmonary blood flow increases.

(C) The mean pressure in the pulmonary arteries rises as cardiac output increases

(D) The pattern of pulmonary blood flow is independent of posture

60. The mammalian respiratory system is characterized by all of the following except :

(A) Alveoli (B) Lamellae

(C) air capillaries (D) Bronchi

61. Tidal volume (VT) in cattle is around :

(A) 500 ml (B) 1 liter (C) 3.5 liter (D) 7.5 liter

62. Effect of oxygen on H+ and Co2 loading and unloading is called :

(A) Boar's effect (B) Haldane's effect

(C) C-D-H effect (D) Both (B) and (C)

63. Reverse chloride shift occurs at :

(A) Tissue level (B) At alveoli

(C) At interstitium (D) Both (A) and (B)

64. About carotid bodies which statement is false :

(A) Are stretch receptors in the walls of the internal carotid arteries

(B) Are influenced more by blood PO_2 than by its oxygen content.

(C) Generate more afferent impulses when blood H^- ion concentration rises.

(D) And the aortic bodies are mainly responsible for the increased ventilation in hypoxia

65. About pulmonary surfactants all are true except :

(A) Increases the surface tension of the fluid lining alveolar walls

(B) Increases lung compliance

(C) Decreases in amount when pulmonary blood flow is interrupted

(D) Increase in fetal lungs during the last month of pregnancy

66. In birds lung surfactant formed a trilaminar substance :

(A) Which prevents transduction of fluid from blood in to air capillary

(B) It is important in preventing collapse of air capillaries

(C) It cover surface of the atria, air capillaries

(D) All of the above

67. Which among the following are ideal condition for maintaining Oxygen-Haemoglobin Dissociation Curve ?

(A) pH 7.4 (B) Temperature 37°C

(C) Concentration of 2,3 BPG 15 uM/g of Hb

(D) All of the above

68. Shifting of Oxygen-Hb Dissociation curve to right indicate :

(A) Increase affinity of Hb for Oxygen

(B) Decrease affinity of Hb for Oxygen

(C) Loading and unloading of Oxygen

(D) Both (A) and (C)

69. Which among the following is not a condition which shifts the curve to right ?

(A) Increase H^+ concentration in blood

(B) Decrease pH

(C) Decrease CO_2 concentration

(D) Increase Temperature

70. In physiological condition shift to right is expected at :

(A) Aorta (B) Arterioles

(C) Tissue capillaries (D) All of the above

71. What is extraction Fraction ?

(A) It is the fraction of atmospheric oxygen taken up in each alveoli

(B) It is the fraction of alveolar oxygen taken up by blood

(C) It is the fraction of oxygen given up by blood at tissue level

(D) Both (B) and (C)

72. Which among the following is false about CO_2 dissociation curve ?

(A) It gives the amount of CO_2 carried by blood in different partial pressure of CO_2

(B) The upper curve is for arterial blood and lower one for venous blood

(C) When the arterial blood becomes venous blood partial pressure of CO_2 changes from 45 mm Hg to 40 mm Hg

(D) 5 unit increase in partial pressure of CO_2 causes 4 volume percent increase in CO_2 transport

73. Which type of respiration occur chiefly in anesthesia ?

(A) Abdominal (B) Thoracic

(C) Both (A) and (B) (D) No respiration

74. During exercise which type of respiration predominates :

(A) Abdominal (B) Thoracic

(C) Both (A) and (B) (D) No respiration

75. Nerve of diaphragm is which of the following :

(A) Thoracic (B) Phrenic (C) Splanchnic (D) Pedal

76. Which of the following is a shallow type of breathing ?

(A) Apnoea (B) Polypnoea (C) Dyspnoea (D) Hypercapnoea

77. Volume change/unit of pressure changed is known as :

(A) Compliance (B) Distensibility

(C) Both (B) and (C) (D) Resistance

78. Which of the following describes the efficiency of breathing ?

(A) Workdone/100 × O_2 Consumed (B) 100/Workdone × O_2 Consumed
(C) Workdone/ O_2 Consumed × 100 (D) None of these

79. In arterial blood, concentration of O_2 and CO_2 respectively are :

(A) 14.4% and 5.5% (B) 20.9 % and 0.03%
(C) 19% and 48% (D) 14% and 52%

80. Alveolar concentration of O_2 and CO_2 respectively are :

(A) 14.4% and 5.5% (B) 20.9 % and 0.03%
(C) 19% and 48% (D) 14% and 52%

81. In arterial blood, concentration of O_2 and CO_2 respectively are :

(A) 14.4% and 5.5% (B) 20.9 % and 0.03%
(C) 19% and 48% (D) 14% and 52%

82. Carbonic anhydrase of RBC have which mineral :

(A) Cu (B) Fe (C) Mn (D) Zn

83. RBC membrane is impermeable to which of the following :

(A) Cation (B) Anion
(C) Both (B) and (C) (D) None of these

84. What is the normal rate of respiration in cow ?

(A) 12 (B) 30 (C) 36 (D) 26

85. Effect of increased O_2 concentration on CO_2 dissolved curve is known as :

(A) Haldane effect (B) CDH effect
(C) Both (A) and (B) (D) Bohr effect

86. In Hb4O8 iron is in .... form.

(A) Fe^{+} (B) Fe^{++}
(C) Fe^{+++} (D) Both (A) and (B)

87. Which of the following is false about metabolic production of acids and bases in our body ?

(A) Metabolism of most of the organic compounds in our body leads to formation of volatile acids
(B) Non-volatile acid like carbonic acid creates less complications
(C) Sulfur containing aminoacid metabolism leads to nonvolatile acid formation like sulfuric acid
(D) Metabolism of feed from plant origin adds excess base to blood stream

88. Which among the following is true ?

(A) Animals eating more protein rich food (meat product) have more acid production
(B) Herbivorous animals have acidic urine whereas carnivorous have alkaline urine
(C) Urine of carnivorous have more potassium
(D) All of the above are true

89. Which one is true about glutamine ?

(A) Ammonium ions formed by protein metabolism used up in hepatic formation of glutamine

(B) Glutamine serves to carry excess hydrogen ions to kidney, where the ions are excreted in the form of ammonium ion
(C) Glutamine plays important role in acid-base balance of the body
(D) All the above are true

90. The relationship between pH and the mixture of weak acid and its conjugate base is defined by :
(A) Isohydric principle
(B) Davenport's diagram
(C) Henderson-Hasselbalch Equation
(D) All of the above

91. Which one is correct Henderson.Hasselbalch Equation ?
(A) pH=K + log (acid/base) (B) pH=pK + log (acid/base)
(C) pH=K + log (base/acid) (D) pH=pK + log (base/acid)

92. Which one among the following is the most important blood buffer ?
(A) Bicarbonate (B) Plasma protein
(C) Phosphate (D) Haemoglobin

93. Which one is false ?
If a strong acid is added to blood in vitro then buffering action by different blood buffers are :
(A) 53% by bicarbonate (B) 25% by haemoglobin
(C) 7% by plasma protein (D) 5% by phosphate

94. Which one among the following is the most important ECF ?
(A) Bicarbonate (B) Protein (C) Phosphate (D) Ammonium

95. What is Isohydric principle ?
(A) It defines the relationship between pH and the mixture of weak acid and its conjugate base
(B) All the buffers present in blood respond at the same time with addition of acid and base
(C) Both are true
(D) None of these

96. In which of the following form CO_2 remain in plasma :
(A) $KHCO_3$ (B) $NaHCO_3$ (C) CO (D) CO_3

97. Formation of carbonic acid in RBC is catalyzed by :
(A) Carbase (B) Carbonase
(C) Carbonic anhydrase (D) Carbonic hydrase

98. Which of the following is false about metabolic acidosis ?
(A) Addition of strong acid or loss of base from the ECF results metabolic acidosis
(B) Diseases like ketosis, diabetes mellitus where acid formation increases results it
(C) Renal acidosis and diarrhea where loss of bicarbonate occur leads to metabolic acidosis
(D) This condition is also called hyperbasemia

99. What are the condition when metabolic alkalosis occur ?
(A) Vomition
(B) Potassium deficiency
(C) Injection or ingestion of salts of organic acids
(D) All of the above

100. Alveolar hyperventilation leads to :
(A) Metabolic acidosis (B) Respiratory acidosis
(C) Metabolic alkalosis (D) Respiratory alkalosis

101. Avian species have different nephron that are :
(A) Cortical and Corticomedullary (B) Mammalian and avian
(C) Reptilian and avian (D) Mammalian and reptilian

102. .... is the major end product of nitrogen metabolism in avian.
(A) Urea (B) Uric acid (C) Ammonia (D) Both (B) and (C)

103. Aquatic animals generally excrete ....
(A) Urea (B) Uric acid (C) Ammonia (D) Both (B) and (C)

104. When birds have salt load they have the ability to conserve water by :
(A) Shifting from reptilian to mammalian nephron
(B) By increasing counter current efficiency
(C) By more protein metabolism
(D) Both (A) and (B)

105. Post renal modification of ureteral urine occurs at :
(A) Colon (B) Caecum (C) Cloaca (D) Ileum

106. Feature unique to birds regarding excretion :
(A) Nephron types are different from mammals
(B) Uric acid is the end product of nitrogen metabolism
(C) In addition to peritubular and glomerular capillaries renal portal system is also present
(D) Counter current system lacks between Henle's loop and medullary collecting duct

107. Juxta-glomerular (JG) apparatus composed of :
(A) JG cells, macula densa and glomerular mesengial cells
(B) JG cells, macula densa and extra-glomerular mesengial cells
(C) JG cells, lacis cells and glomerular mesengial cells
(D) Both (A) and (C)

108. .... marks the beginning of distal tubules in mammalian nephron.
(A) JG cells (B) Macula densa
(C) Lacis cells (D) Mesengial cells

109. Efferent renal sympathetic nerve activities controls :
(A) Ion filtration (B) Renin secretion
(C) Blood flow (D) All of the above

110. JG cells are found in :
(A) Afferent arterioles (B) Efferent arterioles
(C) Thin ascending Henle's loop (D) Thick ascending Henle's loop

111. Macula densa is present at :
(A) Afferent arterioles (B) Efferent arterioles
(C) Thin ascending Henle's loop (D) Thick ascending Henle's loop

112. Rate at which renal plasma is filtered in glomerulus is called :
(A) Renal blood flow (B) Glomerular filtration rate
(C) Filtration fraction (D) Both (B) and (C)

113. Tubular load can be calculated by :
(A) Concentration of substance in plasma x Renal blood flow
(B) Concentration of substance in plasma x glomerular filtration rate
(C) Plasma load x filtration fraction
(D) Both (B) and (C)

114. Which is true about Nitric oxide (NO) :
(A) Potent vasodilator
(B) Produced by vascular endothelial cells
(C) Its substrate is L-arginine
(D) All of the above

115. Myogenic response to stretch is called as :
(A) Latch mechanism (B) Bayliss mechanism
(C) Buffering mechanism (D) Both (B) and (C)

116. Which is false about endothelin ?
(A) A potent vasoconstrictor
(B) Produced by vascular endothelial cells
(C) Consist of 25 amino acids
(D) Found in 3 isoforms

117. Salt glands/nasal glands are present in :
(A) Cattle (B) Sheep (C) Horse (D) Avian

118. About what percent of cardiac output enters into kidney :
(A) 5% (B) 10% (C) 20% (D) 40%

119. Epithelial cells of Bowman's capsule are called :
(A) Proteoglycan (B) Podocyte
(C) Pedicel (D) Slit membrane

120. GFR depends upon :
(A) Blood flow through afferent arterioles
(B) Vasoconstriction of efferent arterioles
(C) Charge and size of substance
(D) All of the above

121. Angiotensinogen is a proenzyme converted to an active form by :
(A) Angiotensin I
(B) Rennin
(C) Renin
(D) Angiotensin converting enzyme (ACE)

122. What percent of sodium ion is reabsorbed at proximal tubule ?
(A) 85 (B) 65 (C) 25 (D) 15

123. Amino acid and glucose molecules are absorbed in proximal tubule by :
(A) Primary active transport (B) Secondary active transport
(C) Facilitated diffusion (D) Diffusion

124. Sodium ion absorption in collecting duct occurs by specialized cells called as :
(A) Intercalated cell (B) Principal cell
(C) Secondary cell (D) Both (B) and (C)

125. Hydrogen ion secretion in collecting duct occurs by specialized cells called as :
(A) Intercalated cell (B) Principal cell
(C) Secondary cell (D) Both (A) and (B)

126. Renal plasma flow (RPF) is measured by :
(A) Inulin (B) Insulin
(C) Mannitol (D) Para amino hippuric acid (PAH)

127. Certain substances exhibit transport maximum (Tm), those :
(A) Are carried by carrier protein (B) Have positive charge
(C) Have size more than 70,000 Dalton (D) All of the above

128. The property of proximal tubules (PT) to reabsorb a constant proportion of filtrate is called :
(A) Tubulo-glomerular balance (B) Glomerulo-tubular balance
(C) Myogenic autoregulation (D) All of the above

129. Kidney's ability to clear a substance from plasma is called :
(A) Renal clearance (B) Tubular clearance
(C) Glomerular clearance (D) All of the above

130. Osmolarity of plasma is about :
(A) 100 mOsm/Kg of water (B) 200 mOsm/Kg of water
(C) 300 mOsm/Kg of water (D) 400 mOsm/Kg of water

131. What is called counter current multiplier :
(A) Loop of Henle (B) Medullary interstitium
(C) Vasa recta (D) All of the above

132. What is called counter current exchanger :
(A) Loop of Henle (B) Medullary interstitium
(C) Vasa recta (D) All of the above

133. ADH (anti diuretic hormone) facilitates large amount of water absorption in :
(A) Thick ascending loop of Henle (B) Distal tubule
(C) Collecting tubule (D) Renal pelvis

134. Urea contributes ...... percent to concentration of urine.
(A) 5 (B) 10 (C) 20 (D) 40

135. pH of urine in herbivorous is usually :
(A) Neutral (B) Alkaline (C) Acidic (D) Both (B) and (C)

136. Dribbling of urine at frequent intervals is called :
(A) Urinary inconsistency (B) Polyuria
(C) Anuria (D) Stranguria

137. Consistency of urine in equines is thick and syrupy due to presence of :
(A) Carbonate (B) Pyruvate (C) Sulphate (D) All of the above

138. Urinary inconsistency is due to :
(A) Stone formation in urinary bladder (B) Spinal injury
(C) Reduced renal blood flow (D) All of the above

139. Responses in one kidney due to interventions of the same or other kidney is called :
(A) Sacro spinal reflex (B) Brain stem reflex
(C) Reno renal reflex (D) None of the above

140. The type of diuretics that inhibit water and solute reabsorption in proximal convoluted tubules is known as :
(A) Loop diuretics (B) Osmotic diuretics
(C) Thiazide diuretics (D) All of the above

141. Increase in urea and other NPN substances due to kidney failure is called :
(A) Urinary inconsistency (B) Uremia
(C) Gout (D) All of the above

142. Specific gravity of urine is :
(A) 1.021 (B) 1.032 (C) 1.045 (D) 1.089

143. Which among the following is an example of carbonic anhydrase inhibitor ?
(A) Furosamide (B) Spironolactone
(C) Mannitol (D) Acetazolamide

144. Which among the following is a loop diuretics ?
(A) Furosamide (B) Spironolactone
(C) Mannitol (D) Acetazolamide

145. What is diuresis ?
(A) Decrease volume of urine output
(B) Increase volume of urine output
(C) Decrease volume of sodium output
(D) Increase volume of sodium output

146. Which among the following is false about regulation of extracellular fluid and electrolyte balance ?
(A) Aldosterone is a regulator of plasma sodium levels
(B) Aldosterone is not a regulator of plasma sodium levels
(C) Aldosterone is not a regulator of plasma potassium levels
(D) Thirst – ADH system is a regulator of plasma sodium levels

147. Increase in sodium and water excretion as a response to increase in blood pressure is called :
(A) Diuresis (B) Pressure natriuresis
(C) Hypovolumia (D) Hypervolumia

148. Angiotensin I is converted to Angiotensin II by :
(A) Renin (B) Rennin
(C) Angiotensin converting enzyme (D) Bradykinin

149. GFR measurement is done with :
(A) Insulin (B) Inulin
(C) Para amino hippuric acid (D) Urea

150. ADH is a nanopeptide produced from :
(A) Parvocellular nucleus of hypothalamus
(B) Supraoptic nucleus
(C) Paraventricular nucleus
(D) Both (B) and (C)

151. Aquaporin, water channels in collecting duct synthesized in response to :
(A) ADH (B) Atrial natriuretic factor
(C) Aldosterone (D) All of the above

152. Odour of cattle urine is faintly aromatic due to :
(A) Calcium phosphate (B) Urobillinogen
(C) Uridine amide (D) Stercobillin

153. Only part of renal tubule in which sodium is secreted :
(A) Ascending thin loop of Henle (B) Descending loop of Henle
(C) Ascending thick loop of Henle (D) Distal tubule

154. Renal medullary blood flow is :
(A) <5% of kidney (B) <10% of kidney
(C) <12% of kidney (D) <15% of kidney

155. Angiotensin converting enzyme (ACE) is produced from :
(A) Vascular endothelium of liver (B) Vascular endothelium of lungs
(C) Vascular endothelium of kidney (D) Both (B) and (C)

156. Organs that does not use glucose insulin for glucose uptake is :
(A) Liver (B) Kidney
(C) Nervous system (D) All of the above

157. Fat and fatty acids are mainly excreted by :
(A) Lungs
(B) Kidney
(C) Skin
(D) Not excreted but all are stored in form of adipose tissue

158. Hydrostatic pressure in renal glomerular capillaries :
(A) Is lower than pressure in efferent arterioles.
(B) Rises when afferent arterioles constrict.
(C) Falls along the length of the capillary.
(D) Same as colloidal osmotic pressure

159. As plasma glucose concentration rises above normal, glucose :
(A) Filtration increases linearly.
(B) Transport maximum Tm increases linearly
(C) Clearance increases linearly
(D) Excretion increases and then decreases

160. Which one is false about renal clearance of a substance ?
(A) Is directly related to its urinary concentration, U
(B) Is directly related to the rate of urine formation, V.
(C) Is directly related to its plasma concentration, P.
(D) Is expressed in units of volume per unit time.

161. Cortical nephrons lack :

(A) Glomerulus (B) Proximal tubule
(C) Descending thin limb (D) Ascending thin limb

162. Blood leaving the glomerulus redistributed in to another capillary bed known as :

(A) Glomerular capillary (B) Peritubular capillary
(C) Vasa recta (D) None of the above

163. Glomerular filtrate after proximal tubule is called as :

(A) Tubular extract (B) Tubular fluid
(C) Renal fluid (D) All of the above

164. Which of the following substances are not filtered during glomerular filtration:

(A) Proteins with molecular weight more than 70,000 Dalton
(B) Amino acids
(C) Salts
(D) Both (A) and (B)

165. Proteoglycans of glomerular basement membrane allow filtration of .... effectively.

(A) Negatively charged particles
(B) Positively charged particles
(C) Proteins
(D) All of the above

166. The intrinsic phenomenon of kidney in order to maintain constant renal blood flow (RBF) and glomerular filtration rate (GFR) with varying systemic arterial pressure.

(A) Tubuloglomerular feedback
(B) Glomerulotubular feedback
(C) Both the above
(D) Medullary washout

167. Important barrier that prevents the free passage of albumin across the glomerular capillary wall is formed by :

(A) Fenestrated capillary endothelium
(B) Anionic proteoglycan molecule
(C) Hydrophilic basement membrane
(D) Filtration slits in between the podocytes

168. Vasopressin increases renal absorption of water fı om :

(A) Proximal tubule (B) Distal tubule
(C) Both the above (D) None of these

169. In hypothyroidism renal phosphate reabsorption :

(A) Decreases
(B) Increases
(C) Remain constant/unaltered
(D) Any of the above can occur/ no relationship

170. During formation of urine in kidney effective filtration pressure in animal is :

(A) 25 mm Hg (B) 35 mm Hg (C) 45 mm Hg (D) 75 mm Hg

171. Which of the following is not a work of kidney ?

(A) Erythropoeisis (B) Bilirubin formation
(C) pH balance (D) Osmotic pressure balance

172. 1,25 Dihydroxy-vitamin D3 is also called :

(A) Calcitriol (B) Calcitonin (C) Calcilock (D) Calmodulin

173. GFR is about .... percent of renal plasma flow.

(A) 5% (B) 10% (C) 20% (D) 30%

174. What is the effect of catacholamines on GFR ?

(A) Decreases (B) Increases
(C) Prevent decrease (D) No effect

175. Which of the following substances that donot exhibit Transport Maximum ?

(A) All substances that are actively transported
(B) Substance that are passively transported
(C) Some actively transported substances that have characteristics of gradient.time transport
(D) Both (B) and (C)

176. Passive water reabsorption is mainly coupled with.......... ion reabsorption.

(A) Sodium (B) Chloride
(C) Calcium (D) All of the above

177. Which among the following is a loop diuretic ?

(A) Furosemide (B) Ethacrynic acid
(C) Bumetanide (D) All of the above

178. What is the role of loop diuretics ?

(A) It increases the Osmolarity of tubular fluid
(B) It inhibits Na^+–K^+–Cl^- co-transport in luminal membrane
(C) Inhibits H^+ secretion
(D) Inhibits action of aldosterone

179. Which among the following is osmotic diuretic ?

(A) Furosemide (B) Acetazolamide
(C) Mannitol (D) Spironolactone

180. Osmotic diuretics mainly act on the :

(A) Proximal tubule (B) Thick ascending loop of Henle
(C) Collecting tubule (D) Distal tubule

— Answer Sheet —

Q. No.	Ans	Q. No.	Ans	Q. No.	Ans	Q. No.	Ans	Q. No.	Ans	Q. No.	Ans
1.	**B**	31.	**B**	61.	**C**	91.	**D**	121.	**C**	151.	**A**
2.	**A**	32.	**C**	62.	**D**	92.	**A**	122.	**B**	152.	**C**
3.	**D**	33.	**B**	63.	**B**	93.	**B**	123.	**B**	153.	**B**
4.	**C**	34.	**C**	64.	**A**	94.	**A**	124.	**B**	154.	**A**
5.	**C**	35.	**D**	65.	**A**	95.	**B**	125.	**A**	155.	**D**
6.	**B**	36.	**C**	66.	**B**	96.	**B**	126.	**D**	156.	**D**
7.	**A**	37.	**D**	67.	**D**	97.	**C**	127.	**A**	157.	**C**
8.	**D**	38.	**C**	68.	**B**	98.	**D**	128.	**B**	158.	**C**
9.	**A**	39.	**A**	69.	**C**	99.	**D**	129.	**A**	159.	**A**
10.	**C**	40.	**A**	70.	**C**	100.	**D**	130.	**C**	160.	**C**
11.	**B**	41.	**D**	71.	**C**	101.	**D**	131.	**A**	161.	**D**
12.	**B**	42.	**D**	72.	**B**	102.	**B**	132.	**C**	162.	**B**
13.	**C**	43.	**D**	73.	**A**	103.	**C**	133.	**C**	163.	**B**
14.	**B**	44.	**C**	74.	**B**	104.	**D**	134.	**D**	164.	**A**
15.	**B**	45.	**B**	75.	**B**	105.	**A**	135.	**B**	165.	**B**
16.	**D**	46.	**B**	76.	**B**	106.	**D**	136.	**A**	166.	**A**
17.	**B**	47.	**C**	77.	**C**	107.	**B**	137.	**A**	167.	**B**
18.	**D**	48.	**B**	78.	**C**	108.	**B**	138.	**B**	168.	**B**
19.	**A**	49.	**B**	79.	**C**	109.	**D**	139.	**C**	169.	**B**
20.	**A**	50.	**D**	80.	**A**	110.	**A**	140.	**B**	170.	**B**
21.	**B**	51.	**D**	81.	**D**	111.	**D**	141.	**B**	171.	**B**
22.	**B**	52.	**C**	82.	**D**	112.	**B**	142.	**B**	172.	**A**
23.	**D**	53.	**A**	83.	**A**	113.	**D**	143.	**D**	173.	**C**
24.	**A**	54.	**D**	84.	**B**	114.	**D**	144.	**A**	174.	**A**
25.	**B**	55.	**B**	85.	**C**	115.	**B**	145.	**B**	175.	**D**
26.	**B**	56.	**C**	86.	**B**	116.	**C**	146.	**A**	176.	**A**
27.	**D**	57.	**D**	87.	**B**	117.	**D**	147.	**B**	177.	**D**
28.	**A**	58.	**A**	88.	**A**	118.	**C**	148.	**C**	178.	**B**
29.	**C**	59.	**C**	89.	**D**	119.	**B**	149.	**B**	179.	**C**
30.	**D**	60.	**B**	90.	**C**	120.	**D**	150.	**D**	180.	**A**

• Chapter 7 •

Muscle Physiology and Nervous System

1. **The basic physiological properties of muscle tissue :**
   (A) Irritability (B) Extensibility
   (C) Elasticity (D) All of the above
2. **Slow twitch muscle is also called as :**
   (A) Type I muscle (B) Red muscle
   (C) Type II muscle (D) Both (A) and (B)
3. **Which statement is false about type II muscle ?**
   (A) High capacity of oxidative phosphorylation
   (B) Produce high amount of force
   (C) Also called as white muscle (D) More fatigable
4. **Which statement is false about type I muscle ?**
   (A) Provide low force
   (B) High level of endurance
   (C) Low myoglobin content
   (D) High capacity of oxidative phosphorylation
5. **Skeletal muscle accounts .... percent of body weight.**
   (A) 20 (B) 40 (C) 60 (D) 80
6. **Regulatory protein in muscle tissue is :**
   (A) Actin and myosin (B) Tropomyosin and troponin
   (C) Titin and myomesin (D) All of the above
7. **Structural protein in muscle tissue is :**
   (A) Actin and myosin (B) Tropomyosin and troponin
   (C) Titin and myomesin (D) All of the above
8. **Troponin protein has .... subunits.**
   (A) 2 (B) 3 (C) 4 (D) 5
9. **Which subunit of troponin binds with myosin ?**
   (A) Troponin I (B) Troponin C (C) Troponin T (D) None
10. **The increased liberation of heat in a stimulated muscle when it is allowed to do mechanical work is called :**
    (A) Latch effect (B) Fenn effect
    (C) Wind kessel effect (D) Latch Mechanism
11. **Size principle for motor unit recruitment is proposed by :**
    (A) Henneman (B) Frankling Starling
    (C) Laplace (D) Kennedy
12. **In Myesthenia gravis :**
    (A) Acetylcholine release is blocked
    (B) Circulating antibodies block acetylcholine receptors at the postsynaptic neuromuscular junction
    (C) More acetylcholine esterase (AchE) release
    (D) All of the above

13. Acetylcholine receptors at the postsynaptic neuromuscular junction are .... in nature.

(A) Leaky channels
(B) Voltage gated channels
(C) Ligand gated channels
(D) Calcium channels

14. Synaptic vesicles are synthesized within................ .

(A) Golgi apparatus
(B) Endoplasmic reticulum
(C) Cytosol
(D) Ribosomes

15. Smooth muscle present in large arteries are called :

(A) Single unit smooth muscle
(B) Multi.unit smooth muscle
(C) Visceral smooth muscle
(D) Unitary smooth muscle

16. Voltage gated calcium channels are present in sarcolemma of smooth muscle, called as :

(A) Dense body (B) Ryanodine
(C) Dihydropyridine (D) Caveolae

17. T tubules are absent in :

(A) Smooth muscle (B) Atrial cardiac muscle
(C) Ventricular cardiac muscle (D) Both (A) and (B)

18. Calcium induced calcium release is a feature unique for :

(A) Skeletal muscle (B) Cardiac muscle
(C) Smooth muscle (D) Both (A) and (B)

19. Each muscle twitch has .... phase.

(A) 2 (B) 3 (C) 4 (D) 5

20. Primary energy source for cardiac muscle is :

(A) Creatine phosphate (B) Glucose
(C) Fatty acid (D) Lactate

21. The contractile protein of skeletal muscle involving ATPase activity is :

(A) Actin (B) Myosin
(C) Troponin (D) Tropomyosin

22. Which of the following is the contractile protein of a muscle ?

(A) Tubulin (B) Myosin
(C) Tubulin (D) All of these

23. Muscle fatigue results in due to nonavailability of :

(A) Calcium (B) ATP
(C) actin binding site (D) Mg cofactor

24. Muscles get fatigue due to accumulation of :

(A) Lactate (B) Phosphate
(C) Carbon dioxide (D) ATP

25. Which one of the following sets of ions are necessary in the chemical events for muscle contraction?

(A) Ca^{2+} and Mg^{2+} (B) Na^{+} and Ca^{2+}
(C) Na^{+} and K^{+} (D) Na^{+} and Ca^{2+}

26. Upon stimulation of skeletal muscles calcium is immediately made available for binding to troponin from :

(A) blood (B) sarcoplasmic reticulum
(C) lymph (D) Both (B) and (C)

27. Choose one correct answer :

(A) Cardiac muscle requires beat-to-beat Ca influx to trigger contraction.
(B) A DHP blocker will increase the strength of contraction in heart muscle
(C) SR Ca release channels are called DHP receptors
(D) None are correct

28. Choose one correct answer :

(A) Skeletal muscle fibers that rely mainly on oxidative phosphorylation for ATP production have few capillaries
(B) Creatine kinase catalyzes the phosphorylation of creatine phosphate
(C) Heart muscle cells can normally develop tetanic contractions.
(D) The calcium–calmodulin complex in smooth muscle inhibits myosin-light chain kinase

29. Endurance training increases .... in skeletal muscle cells.

(A) The capillary density
(B) Enzyme involved in citric acid cycle
(C) The size and number of mitrocondria
(D) All of the above

30. Muscle cell hyperplasia is an adaptive mechanism can be observed in all muscle type except :

(A) Skeletal muscle
(B) Cardiac muscle
(C) Smooth muscle
(D) White muscle

31. Which among following is false about smooth muscle ?

(A) Considered as primitive form of muscle tissue.
(B) Lacks actin and myosin protein, therefore it is non.striated
(C) Smooth muscle tissue has much more extracellular material than other muscle tissue types
(D) Wall of large elastic arteries are made up off abundant smooth muscle

32. Smooth muscle cells of .... shape.

(A) Star shape (B) Cylindrical shape
(C) Spindle shape (D) Circular shape

33. In smooth muscle actin fillaments are anchored with :

(A) Z line (B) M line
(C) Dense bodies (D) Caveolae

34. Which among following is false about smooth muscle ?

(A) The calcium-calmodulin complex inactivates myosin light chain kinase
(B) Sarcoplasmic reticulum is rudimentary in smooth muscle
(C) Myosin ATPase activity is much slower in smooth muscle
(D) Hormone-activated calcium ion channels are present on cell membrane of smooth muscle

35. Which among following is false about smooth muscle ?

(A) T tubules are absent in smooth muscle cells
(B) Extracellular calcium has no influence in smooth muscle contraction
(C) Calcium pump present on the cell membrane of smooth muscle are much sslower than sarcoplasmic reticulum of smooth muscle
(D) Calmodulin is present in smooth muscle cell

36. Which among the following is true about multi-unit smooth muscle ?

(A) In this smooth muscle type large area of muscle tissue contract simultaneously
(B) Gap junctions are present
(C) Autorhythmicity is a phenomenon unique to this muscle type
(D) Found in ciliary body and iris of eye

37. Smooth muscle is able to stay contracted for some time without the use of much ATP called :

(A) Contracted state (B) Latch state
(C) Kinase state (D) Active state

38. Usual cause of tetany is :

(A) Lack of calcium or excess of phosphate
(B) Excess of calcium or lack of phosphate
(C) No role of calcium and phosphate
(D) Excess of lactic acid

39. Which statement is false about muscle tetany ?

(A) A common cause of tetany is hypocalcemia
(B) Involuntary contraction of muscles
(C) Tetany can affect any muscle in the body
(D) Tetany refers to a state of decreased excitability of nerve and muscle

40. Which is false about size principle ?

(A) Proposed by Henneman
(B) Motor units are generally recruited in order of smallest to largest
(C) It minimizes the amount of fatigue an organism experiences by using fatigue.resistant muscle fibers first and only using fatigable fibers when high forces are needed
(D) There exists no force-frequency relationship

41. Calmodulin is involved in :

(A) Smooth muscle relaxation (B) Skeletal muscle contraction
(C) Smooth muscle contraction (D) Skeletal muscle relaxation

42. Regarding smooth muscles ignition of contraction involves which mechanism

(A) Na influx (B) Ca influx (C) K influx (D) Na efflux

43. Which is true regarding cardiac muscle ?

(A) Gap junctions resist the flow of ions
(B) Like skeletal muscle multinucleated cells are present
(C) Its intercalated discs are loosely attached
(D) Z line is present like skeletal muscle

44. Which is correct about skeletal muscle ?

(A) The immediate energy source of contraction is GTP
(B) Troponin T inhibits the interaction of myosin with actin
(C) Tropomyosin is made up off 3 sub-units
(D) Myosin is contained entirely within the A band

45. Which of the following is the special feature of smooth muscle contraction ?

(A) Actin is not involved (B) Myosin is not involved
(C) ATP is not the energy source (D) The membrane potential is unstable

46. When skeletal muscle contracts :

(A) The H zone increases (B) The A and I zone increases
(C) The I zone decreases (D) The A zone decreases

47. Thick filament of skeletal muscle is composed of :

(A) Actin (B) Myosin (C) Troponin (D) Calmodulin

48. Which of the following statements regarding the shortening of a skeletal-muscle fibre is not true? When a skeletal-muscle fibre shortens ?

(A) the sarcomeres shorten
(B) the distance between Z lines decreases
(C) the myofilaments shorten
(D) the length of the A bands remains the same

49. In skeletal muscle, calcium facilitates contraction by binding to :

(A) Tropomyosin (B) Troponin (C) Actin (D) Myosin

50. Rigor mortis occurs in a dead animal because :

(A) ATP, which is necessary for the detachment of cross bridges, is not being formed
(B) ATP, which is necessary for the formation of cross bridges, is not being formed
(C) ATP, which is necessary for the formation of cross bridges, continues to be formed for several hours after death
(D) deterioration of muscle proteins prevents detachment of cross bridges

51. An action potential in the motor end plate rapidly spreads to the central portions of a muscle cell by means of the :

(A) Transverse tubule (B) Sarcoplasmic reticulum
(C) Pores of cell membrane (D) Z line

52. According to the sliding filament mechanism of skeletal.muscle contraction, during contraction :

(A) The thick filaments stay the same size but the thin filaments shorten
(B) The sarcomeres shorten
(C) The thin filaments stay the same size but the thick filaments shorten
(D) Both thick and thin filaments shorten

53. Gap junctions of intercalated disc contains .... protein.
(A) Cladherin (B) Ocludin (C) Connexon (D) Desmosin

54. Which among the following is true about cardiac muscle ?
(A) Cardiac muscle cells behave like a functional syncytium
(B) Mitochondria makeup about 40% of cytoplasmic volume
(C) T.tubules are poorly developed in arterial in atrial cardiac muscle
(D) All of the above

55. Skeletal muscle fiber is :
(A) Anucleated (B) Mononucleated
(C) Binucleated (D) Multinucleated

56. "Motor unit" refers to :
(A) a single motor neuron plus all the muscle fibres it innervates
(B) a single muscle fibre plus all of the motor neurons that innervate it
(C) all of the motor neurons supplying a single muscle
(D) all of the muscles that affect the movement of any given joint

57. During an isometric contraction of a skeletal muscle :
(A) the I bands shorten and the A bands stay the same length
(B) the thick and thin filaments slide past each other
(C) sarcomere length does not change
(D) Both (A) and (C)

58. Which of the following is not true regarding the comparison of type I (slow oxidative) and type II b (fast.glycolytic) skeletal.muscle fibres ?
(A) Type I fibres have more abundant mitochondria
(B) Type I fibres fatigue more readily
(C) Type I fibres have more abundant myoglobin
(D) Type I fibres have more abundant capillaries

59. Cardiac muscle is different from skeletal muscle because :
(A) Fast Na Channels (B) Slow Ca Channels
(C) Presence of actin and myosin (D) Presence of troponin

60. Tetany does NOT occur in cardiac muscle because :
(A) Long absolute refractory period
(B) Acts as a syncytium
(C) Pacemaker signal can overcome any tetany
(D) All are true

61. An increase in force of a skeletal muscle contraction is initially achieved by :
(A) Recruitment of nerve fibres (B) Recruitment of muscle fibres
(C) Recruitment of motor units (D) All are true

62. Calcium ions are released from the sarcoplasmic reticulum to initiate muscle contraction and are reabsorbed upon muscle relaxation. Which of the following occurs if calcium ions are not completely reabsorbed ?
(A) The next muscle contraction will be greater than the previous muscle contraction
(B) The next muscle contraction will be weaker than the previous muscle contraction

(C) There won't be a second muscle contraction until all the calcium ions are reabsorbed
(D) Any consequence may happen

63. Which of the following is the correct sequence for the production of ATP for muscle contraction?
(A) anaerobic respiration, CP, aerobic respiration
(B) aerobic respiration, CP, anaerobic respiration
(C) CP, anaerobic respiration, aerobic respiration
(D) Anaerobic respiration, aerobic respiration, CP

64. ATP is required for muscles to contract. Which of the following statements is FALSE ?
(A) ATP is used to get the cross–bridges to bind to actin
(B) ATP is used to get the cross–bridges to disconnect from actin
(C) ATP is used to get the troponin–tropomyosin complex to move in such a manner to expose the binding sites
(D) Both (A) and (C)

65. ..... surrounds each muscle fasciculus, whereas ..... surrounds each muscle fiber.
(A) Epimysium, perimysium (B) Endomysium, perimysium
(C) Perimysium, epimysium (D) Perimysium, endomysium

66. Cross bridges form during muscle contraction when myosin comes in contact with active sites on :
(A) Troponin molecule (B) Tropomyosin molecule
(C) Calcium ions (D) Actin molecule

67. Which of these is (are) required for the relaxation of a muscle fiber ?
(A) ATP
(B) Ca^{+2} ion uptake by sarcoplasmic reticulum
(C) Acetylcholinesterase activity
(D) All of the above

68. A muscle contraction in which the muscle produces an increasing tension, but the length remains constant is called :
(A) Isometric contraction (B) Isotonic contraction
(C) Trepp (D) Concentric contraction

69. Fatty acids can be used as an energy source for muscle contraction during :
(A) Aerobic exercise (B) Anaerobic respiration
(C) Fast and intense exercise (D) (B) and (C)

70. Which of these regions shorten(s) during skeletal muscle contraction ?
(A) A band (B) I band (C) H band (D) Both (B) and (C)

71. A type of interneuron found in the ventral horn of spinal cord involved in the recurrent inhibition phenomenon is called :
(A) alpha motor nerve (B) gamma motor nerve
(C) Renshaw cell (D) Contra lateral nerve

72. Which of the following proteins has ATPase activity ?
(A) Troponin (B) Tropomyosin
(C) Actin (D) Myosin

73. An Ach receptor at a neuromuscular unit contains following 5 subunits :

(A) 2α, 2β, 1γ (B) 2α, 1β, 1γ, 1δ

(C) 1α, 2β, 1γ, 1δ (D) 2α, 1β ,2γ

74. Maximal contraction occurs in a skeletal muscle, when there is :

(A) Maximal overlapping of actin over myosin

(B) Minimal overlapping of actin over myosin

(C) Overlapping of actin is up to half of myosin

(D) None is correct

75. Which one is false about Rigor mortis ?

(A) Permanent hardening of body after death

(B) Under normal condition lasts for 15-25 hours

(C) Muscle rigidity is lost rapidly under cold climate

(D) Occurs due to lack of ATP

76. Which of the following statement is true about neuromuscular junction ?

(A) Motor end plates are covered by Schwann cells

(B) Nerve invaginates muscle fiber at its periphery

(C) Usually there are many neuromuscular junction in a single muscle fiber

(D) The invaginating nerve fiber penetrate the plasma membrane of muscle fiber

77. Which statement is false about acetylcholine (Ach) receptor of neuromuscular junction ?

(A) Is a metabotropic receptor

(B) It allows passage of Na^+ ion when in open state

(C) It allows passage of Cl^- ion when in open state

(D) It opens when Ach binds to α subunit

78. The action potential of skeletal muscle :

(A) spreads inwards to all parts of the muscle via T tubules

(B) has a prolonged plateau phase

(C) is not essential for muscle conduction

(D) is longer than the action potential of cardiac muscle

79. The likely mechanism through which neostigmine acts in improving muscular weakness is :

(A) It blocks action of acetylcholine

(B) it blocks the action of acetyl choline esterase

(C) it enhances the action of catecholamines

(D) it interferes with action of mono.amine oxidase

80. The function of tropomyosin in skeletal muscle is :

(A) sliding on actin to produce shortening

(B) Releasing Ca^{++} after initiation of contraction

(C) binding to myosin during contraction

(D) covering up the actin binding sites of myosin at rest

81. Regarding phosphocreatine which of the following is true :

(A) It is synthesized in liver

(B) It is excreted in urine

(C) During exercise it reacts with ADP
(D) Rate of phosphocreatine excretory product remains constant throughout life.

82. **The immediate source of energy for skeletal muscle contraction is :**
(A) ATP (B) Glucose
(C) Creatine phosphate (D) Fatty acid

83. **Cross bridges are formed at which part of sarcomere :**
(A) H zone (B) Z line (C) A zone (D) I band

84. **The time required by the muscle from the point of stimulus to the beginning of contraction is called :**
(A) Contraction period (B) Refractory period
(C) Latency period (D) Relaxation period

85. **When a second stimulus is applied quickly after first stimulus there is no response to second stimulus this period is called :**
(A) Contraction period (B) Refractory period
(C) Latency period (D) Relaxation period

86. **If repetitive stimuli are applied to muscle with long periods of interval. Individual contraction can be seen. This condition is known as :**
(A) Incomplete tetanus (B) Complete tetanus
(C) Treppe (D) Fatigue

87. **Onset of rigor mortis is directly proportional to amount of :**
(A) Glucose (B) Glycogen (C) Galactose (D) All of the above

88. **Tropomyosin is a :**
(A) Contractile protein (B) Regulatory protein
(C) Lipoprotein (D) None of these

89. **Which part derived from ectoderm :**
(A) Cardiac muscle (B) Gonad (C) Tonsils (D) Kidney

90. **The notochord is replaced by :**
(A) Vertebral canal (B) Spinal cord
(C) Cerebellum (D) Dorsal root ganglia

91. **Development and differentiation appears first in :**
(A) Head (B) Tail
(C) Limb (D) In all organs at a time

92. **Earliest gland to apper in embryo is :**
(A) Pituitary (B) Thyroid (C) Adrenal (D) Testis

93. **Discovery of ovum is by :**
(A) De Graaf (B) Von Baer (C) Turner (D) Gullyman

94. **The nerve cells of autonomic gaglia are mainly :**
(A) Pseudounipolar (B) Unipolar
(C) Bipolar (D) Multipolar

95. **Fourth ventricle of brain is :**
(A) Diencephalon (B) Telencephalon
(C) Mylencephalon (D) Metencephalon

96. The portion of cell body from where the axon originates :

(A) Soma (B) Axon Hillock
(C) Trigger Zone (D) Both (B) and (C)

97. The primary receptive site of impulses in a neuron is :

(A) Soma (B) Axon Hillock
(C) Dendrites (D) Neurolemma

98. The axon potential is originated at :

(A) Soma (B) Axon Hillock
(C) Dendrites (D) Neurolemma

99. Which statement about neurons is false ?

(A) Myelinated fibers carry impulses slower than Unmyelinated fibers
(B) Neurons communicate with each other via synapses
(C) Areas of brain containing myelinated fibers appears grey
(D) Myelin are made up of lipids

100. Which types of neurons are present in retina of eye ?

(A) Unipolar neuron (B) Bipolar neuron
(C) Multipolar neuron (D) Any one may be there

101. Which statement is false about Schwann cell ?

(A) These are a type of neuroglia cell
(B) These are found in peripheral nervous system
(C) Schwann cells are arranged side by side along the axon
(D) These are the smallest neuroglia cell

102. Which one is false about microglia ?

(A) These are the macrophages of CNS
(B) Microglia are known to produce Nitric oxide
(C) 10-20% of all neuroglia is microglia
(D) Most abundant neuroglia is microglia

103. The astrocytic processes that cover capillaries are called :

(A) Neurotropic factor (B) Perivascular end feet
(C) Extracellular adhesion molecules (D) Limiting membrane

104. Ventricles and central canal of spinal cord are lined by :

(A) Schwann cells (B) Ependymal cells
(C) Astrocytes (D) Oligodendrocytes

105. Acetylcholine is synthesized in :

(A) Golgi apparatus (B) Mitrochondria
(C) Ribosomes (D) All of the above

106. The resting membrane potential of most membranes is :

(A) –70 mV (B) +90 mV (C) –80mV (D) –90mV

107. Every ion has a :

(A) Resting potential (B) Membrane potential
(C) Equilibrium potential (D) Reverse potential

108. Equillibrium potential of calcium ion is :

(A) –70 mV (B) +90 mV (C) +70mV (D) –90mV

109. Equillibrium potential of sodium ion is :
(A) –70 mV (B) +90 mV (C) +70mV (D) –90mV

110. Rapid reversal of polarity towards positive side is called :
(A) After potential (B) Repolarization
(C) Depolarization (D) Hyperpolarization

111. Various pre synaptic potentials are added up at :
(A) Dendrites (B) Axon
(C) Axon hillock (D) Soma

112. Repolarization also called as :
(A) Overshoot of Action potential (B) Spike potential
(C) Falling phase (D) Depolarization

113. Myelin is formed by :
(A) Glial cell (B) Nerve cells
(C) Axon membrane (D) All of the above

114. Which statement is true about myelin sheath ?
(A) Myelin sheath surrounds all the axon of the nervous system
(B) Myelin sheath is continuous throughout the axon without interruption
(C) It act as an insulator and helps in fast movement of impulse
(D) All the above are true

115. The speed with nerve impulse travel depends on :
(A) Diameter of nerve (B) Myelination
(C) Both (A) and (B) (D) Type of impulse

116. Dorsal root ganglia are example of :
(A) Pseudounipolar neuron (B) Monopolar neuron
(C) Bipolar neuron (D) Tripolar neuron

117. Which among the following is the function of glial cell ?
(A) They release cytokine during disease process
(B) They don't participate directly in propagation of nerve impulse
(C) They form blood brain barrier
(D) All the above are true

118. Macroglia includes :
(A) Astrocytes (B) Oligodendrocytes
(C) Ependymal cell (D) Both (A) and (B)

119. Which type of glial cell form blood brain barrier ?
(A) Astrocytes (B) Oligodendrocytes
(C) Ependymal cell (D) Microglia

120. Resident macrophage of CNS are :
(A) Astrocytes (B) Oligodendrocytes
(C) Ependymal cell (D) Microglia

121. When an axon is interrupted or cut the distal end of the same degenerates it is called :
(A) Programmed cell death (B) Wallerian degeneration
(C) Both the above (D) None of these

122. Schwann cells of peripheral nervous system correspond to .... of CNS.

(A) Astrocytes (B) Oligodendrocytes
(C) Ependymal cell (D) Microglia

123. Identify the correct matching of brain structure and the part of the neural tube from which it derives :

(A) Hypothalamus - diencephalon
(B) Frontal cortex - Prosencephalon
(C) Cranial nerves IVI to XII - Myelencephalon
(D) Cerebellum - mesencephalon

124. General somatic efferent fibers are found in all the following cranial nerves except :

(A) II (B) III (C) IV (D) V

125. The central and peripheral nervous system are derived from :

(A) Ectoderm (B) Mesoderm
(C) Both (A) and (B) (D) Endoderm

126. Which statement is true about ion channels ?

(A) Usually ion channels are selective, allowing one ion
(B) Ion channels commonly composed of four or five protein subunits
(C) Ion channels are mainly two types i.e. leakage and gated channel
(D) All the above are true

127. Which one is false about ion channels ?

(A) These are classified as voltage and chemically gated
(B) Open and closed state of gated channel depends upon the presence of stimuli
(C) Leakage channels also have open and closed state which depend upon presence of ions
(D) The subunits of ion channels generally span the lipid bilayer

128. What is false about action potential :

(A) It is small and slow change in membrane potential
(B) This is the signal that the nervous system uses to transfer information over long distance
(C) It occurs on the axon
(D) It is also called an all or none phenomenon

129. What is true about postsynaptic potential :

(A) Smaller changes in membrane potential
(B) Change in membrane potential may be positive or negative
(C) It usually occurs in cell body or dendrites
(D) They can travel long distance

130. Which one is not true about both action potential and post synaptic potential ?

(A) Action potential is rapid and fast where as PSP is slow and small change in membrane potential
(B) Action potential is always positive where as PSP may be positive or negative

(C) Action potential is by voltage gated ion channel whereas PSP by chemically gated channels
(D) Action potential and PSP both once initiated it travels the whole neuron i.e. called all or none phenomenon

131. Slow repolarisation is also called :
(A) Depolarization phase (B) Resting phase
(C) Negative after potential (D) Positive after potential

132. Hyperpolarization occurs due to increase in permeability of :
(A) Sodium ion (B) Calcium ion
(C) Potassium ion (D) Chloride ion

133. Hyperpolarization is also called :
(A) Resting phase (B) Rising phase
(C) Falling phase (D) Recovery phase

134. What is triggering level or threshold potential to achieve action potential :
(A) –30 to –20mV (B) –20 to 0 mV
(C) 0 to 10mV (D) 10 to 15 mV

135. Leaky channels are mainly selective for :
(A) Sodiumion (B) Calciumion
(C) Potassiumion (D) Chlorideion

136. Different states of sodium voltage gated channels are :
(A) Open state and close state (B) Open state and inactivation state
(C) Inactivation state and close state (D) Open, inactivation and close state

137. When the sodium voltage gated channels are in inactivation state then it is impossible to initiate another action potential called as :
(A) Absolute refractory period (B) Relative refractory period
(C) Long QT syndrome (D) Negative after potential

138. During repolarization phase sodium gated channels gradually shift from inactivation to closed state, at this time greater amount of depolarization is needed to initiate an action potential :
(A) Absolute refractory period (B) Relative refractory period
(C) Long QT syndrome (D) Negative after potential

139. In heart muscle mutation in cardiac sodium channels can contribute to electrical dysfunction is a disorder called as :
(A) Absolute refractory period (B) Relative refractory period
(C) Long QT syndrome (D) Negative after potential

140. Once action potential generated it propagate through the axon through :
(A) Voltage gated channels (B) Electronic current spread
(C) Both (A) and (B) (D) None of these

141. Due to the presence of Nodes of Ranvier in myelinated axons and due to electronic current it appears as if action potential leap from node to node, this type of conduction is called :
(A) Saltatory conduction (B) Internodal conduction
(C) Unidirectional conduction (D) Passive conduction

142. For release of neurotransmitter which ion is essential :

(A) Sodium (B) Potassium
(C) Calcium (D) Chloride

143. Which among the following is not an event involved in neurotransmitter release?

(A) Pumping (B) Mobilization
(C) Docking (D) Fusion

144. Synaptic vesicles are attached to cytoskeletal structures by a protein called as :

(A) SNAP 25 (B) Synapsin I
(C) Syntaxin (D) Synaptotagmine

145. Calcium-Calmodulin complex modifies the synapsin I protein and leading to detachment of detachment of vesicles and make them free to move towards synaptic boutonscalled as :

(A) Release (B) Mobilization
(C) Docking (D) Fusion

146. The anchoring of vesicular membrane binding protein synaptobrevin with synaptic bouton membrane protein syntaxin and SNAP-25 is called :

(A) Release (B) Mobilization
(C) Docking (D) Fusion

147. Which vesicular membrane protein act as calcium sensor ?

(A) SNAP 25 (B) Synapsin I
(C) Syntaxin (D) Synaptotagmine

148. Which protein play crucial role in endocytosis or recycling of neurotransmitter vesicle ?

(A) SNAP 25 (B) Synapsin I
(C) Syntaxin (D) Synaptotagmine

149. Which among the following is true about ionotropic receptors ?

(A) It is a class of chemically gated channels
(B) Made up off several subunits spanning across the membrane
(C) Nicotinic acetylcholine receptor of skeletal muscle are example of ionotropic receptor
(D) All the above are true

150. Which among the following is false about Nicotinic acetylcholine receptor of skeletal muscle ?

(A) It is a class of ionotropic receptors
(B) It has 4 protein subunits (2 α and 2 β)
(C) Both the α subunits have a high affinity for extracellular binding of acetylcholine
(D) Opening of this ion channel permits the flow of sodium, potassium and calcium across the membrane

151. Which among the following is true about metabotropic receptors ?

(A) It is a class of chemically gated channels
(B) Single membrane spanning unit
(C) G protein coupled receptors are example of it
(D) All the above are true

152. When a second postsynaptic potential (PSP) generates before an initial PSP is fully decayed, resulting in a partial summation called as :
(A) United PSP (B) Spatial summation
(C) Temporal summation (D) All the above are true

153. Which among the following is not a small molecule neurotransmitter ?
(A) Glutamate (B) Oxytocin (C) Dopamine (D) Glycine

154. Mono amine oxidase (MAO) is used in the catabolism of :
(A) Catecholamines (B) Acetylcholine
(C) Both (A) and (B) (D) None of these

155. Which among the following helps in catabolism of catecholamines ?
(A) Acetylcholinesterase
(B) Mono amine oxidase (MAO)
(C) Catechol O methyl transferease (COMT)
(D) Both (B) and (C)

156. Referred pain is felt on the surface of body but the sensation of this arise from:
(A) Deep structures (B) Body surface
(C) Both (A) and (B) (D) No actual pain

157. Once the membrane potential has returned to its resting levels, it becomes a little more negative than its normal resting value, the excess negative is called :
(A) Positive after potential (B) Negative after potential
(C) Threshold potential (D) Repolarization

— Answer Sheet —

Q. No.	Ans	Q. No.	Ans	Q. No.	Ans	Q. No.	Ans	Q. No.	Ans	Q. No.	Ans
1.	**D**	21.	**B**	41.	**C**	61.	**C**	81.	**C**	101.	**D**
2.	**D**	22.	**B**	42.	**B**	62.	**A**	82.	**A**	102.	**D**
3.	**A**	23.	**B**	43.	**D**	63.	**C**	83.	**C**	103.	**B**
4.	**C**	24.	**A**	44.	**D**	64.	**C**	84.	**C**	104.	**B**
5.	**B**	25.	**A**	45.	**D**	65.	**D**	85.	**B**	105.	**B**
6.	**B**	26.	**B**	46.	**C**	66.	**D**	86.	**A**	106.	**A**
7.	**C**	27.	**A**	47.	**B**	67.	**D**	87.	**B**	107.	**C**
8.	**B**	28.	**B**	48.	**C**	68.	**A**	88.	**B**	108.	**D**
9.	**D**	29.	**D**	49.	**B**	69.	**A**	89.	**C**	109.	**C**
10.	**B**	30.	**B**	50.	**A**	70.	**D**	90.	**A**	110.	**C**
11.	**A**	31.	**B**	51.	**A**	71.	**C**	91.	**A**	111.	**C**
12.	**B**	32.	**C**	52.	**B**	72.	**D**	92.	**B**	112.	**C**
13.	**C**	33.	**C**	53.	**C**	73.	**B**	93.	**B**	113.	**A**
14.	**A**	34.	**A**	54.	**D**	74.	**A**	94.	**D**	114.	**C**
15.	**B**	35.	**B**	55.	**D**	75.	**C**	95.	**C**	115.	**C**
16.	**D**	36.	**D**	56.	**A**	76.	**A**	96.	**D**	116.	**A**
17.	**D**	37.	**B**	57.	**C**	77.	**C**	97.	**C**	117.	**D**
18.	**B**	38.	**A**	58.	**B**	78.	**A**	98.	**B**	118.	**D**
19.	**B**	39.	**D**	59.	**B**	79.	**B**	99.	**C**	119.	**A**
20.	**C**	40.	**D**	60.	**A**	80.	**D**	100.	**B**	120.	**D**

121. **B**	127. **A**	134. **D**	141. **A**	148. **D**	155. **D**
122. **B**	128. **A**	135. **C**	142. **C**	149. **D**	156. **A**
123. **D**	129. **D**	136. **D**	143. **A**	150. **B**	157. **A**
124. **A**	130. **D**	137. **A**	144. **B**	151. **D**	
125. **A**	131. **C**	138. **B**	145. **B**	152. **C**	
126. **D**	132. **C**	139. **C**	146. **C**	153. **B**	
127. **C**	133. **D**	140. **B**	147. **D**	154. **A**	

• Chapter 8 •

Reproductive System

1. .... region of genital ridge develop into testes.

(A) Cortex (B) Medulla
(C) Mesenchymal cell (D) Wolffian duct

2. Gonadotropins exert their negative feedback at .... nucleus of hypothalamus.

(A) Arcuate (B) Antero-ventral periventricular
(C) Preoptic (D) All of the above

3. Sigmoid flexor is absent in :

(A) Cattle (B) Pig (C) Horse (D) Both (B) and (C)

4. Among the following accessory sex gland in dog .... is present namely.

(A) Seminal vesicle (B) Prostate
(C) Bulbourethral (D) Both (A) and (B)

5. Ovary is .... shape in cows.

(A) Almond (B) Bean (C) Mulberry (D) Oval

6. In stallion testosterone is produced by :

(A) Leydig cell (B) Sertoli cell (C) Epididymis (D) Both (A) and (C)

7. In horse and pig for successful insemination semen need to be deposited.

(A) In caudal end of vagina (B) In cranial end of vagina
(C) In caudal end of cervix (D) In cranial end of cervix

8. Prostate gland is only discrete type in :

(A) Horse (B) Bull (C) Boar (D) Dog

9. Presence of protoplasmic droplet is :

(A) Is a normal sign (B) Is a sign of infertility
(C) Is a sign of immaturity (D) Is a sign of end of spermatogenesis

10. Process of release of germ cells in to the lumen of seminiferous tubule is called :

(A) Spermatogenesis (B) Spermiogenesis
(C) Spermateliosis (D) Spermiation

11. Largest testis in proportion to body weight is in :

(A) Bull (B) Boar (C) Stallion (D) Dog

12. Principal permeability barrier of blood-testis barrier is formed by :

(A) Myoid cell (B) Sertoli cell (C) Leydig cell (D) Both (A) and (B)

13. Orientation of testis is vertical in :

(A) Bull (B) Boar (C) Stallion (D) Dog

14. .... in the seminal plasma primarily maintains the osmotic pressure.

(A) Ergothionine (B) Citric acid (C) Inositol (D) Fructose

15. A term describing reduced sperm motility :

(A) Asthenozoospermia (B) Oligozoospermia
(C) Necrozoospermia (D) Hypospermia

16. Activin is structurally :

(A) $1\alpha+1\beta$ subunit of Inhibin
(B) 2α subunit of Inhibin
(C) 2β subunit of Inhibin
(D) Similar to Inhibin with an additional glucose moiety

17. Quiescence factor in bull is secreted by :

(A) Seminal vesicle (B) Cowper's gland
(C) Epididymis (D) Sertoli cell

18. Only accessory gland common to all animals is :

(A) Seminal vesicle (B) Ampula
(C) Prostate (D) Bulbourethral

19. Oocytes are arrested at .... stage upto birth called Dictyate stage.

(A) Prophase of meiosis I (B) Prophase of meiosis II
(C) Metaphase of meiosis I (D) Metaphase of meiosis II

20. Corpus luteum of pregnancy is called :

(A) Corpus luteum spurium (B) Corpus luteum vernum
(C) Corpus luteum albican (D) Corpus luteum haemorragicum

21. Blood flow to ovary is maximum at :

(A) Follicular phase (B) Ovulation
(C) Luteal phase (D) Both (B) and (C)

22. In which species the presence of corpus luteum is not a bare requirement for maintainance of pregnancy :

(A) Cow (B) Doe (C) Mare (D) Sow

23. Fertilization in most mammalian species takes place at :

(A) Infundibulum (B) Ampulla
(C) Ampuloisthmic junction (D) Isthmus

24. Sperm capacitation occurs when sperm travels trough :

(A) Epididymis
(B) Cervix of female reproductive tract
(C) Uterus of female reproductive tract
(D) In cervical mucus at the time of colonization

25. Gonadotropin releasing hormone is a :

(A) Tripeptide (B) Pentapeptide (C) Nanopeptide (D) Decapeptide

26. The credit of first successful cryopreservation in cattle embryo goes to :

(A) Whittingham et al., 1979 (B) Audrey and Smith, 1952
(C) Wilmut and Rowson, 1973 (D) Trounson et al., 1982

27. The credit of first successful cryopreservation in sheep embryo goes to :

(A) Bilton and Moore, 1976 (B) Bilton and Moore, 1979
(C) Willadsen et al., 1976 (D) Wilmut and Rowson, 1973

28. During the process of cryopreservation solution effect is seen as a result of :

(A) Ultra-fast freezing (B) Fast freezing
(C) Slow freezing (D) Both (A) and (B)

29. Cryoprotectants like glycerol and DMSO results :
(A) Freezing at lower temperature (B) Retard dehydration of cells
(C) Prevent large ice crystal formation
(D) Both (A) and (B)

30. Minimum standard for classification of probably fertile bull semen is :
(A) Over 500 million spermatozoa per ml
(B) > 50% of motile sperm makes forward progression
(C) > 80% of spermatozoa conform to normal morphology
(D) All of the above

31. CASA calculates :
(A) Sperm cell concentration
(B) Concentration of motile sperm
(C) Average linearity of forward progression
(D) All of the above

32. Asthenozoosprmia indicates :
(A) Zero sperm volume (B) Decreased sperm volume
(C) Zero sperm concentration (D) Decreased sperm motility

33. Hypo-osmotic swelling test is used for :
(A) Evaluating the morphology of spermatozoa
(B) Evaluating the functional integrity of spermatozoa
(C) Segregation of sexed semen
(D) Both (A) and (B)

34. In Hamster test :
(A) Hamster ova with Zona pellucida is used
(B) Zona free Hamster ova is used
(C) Capacitated Hamster sperm is used
(D) Fresh ejaculated Hamster sperm is used

35. Tertiary abnormality in spermatozoa is due to :
(A) Failure of spermatogenesis
(B) Abnormalities occur during passage of spermatozoa through epididymis
(C) Damage of spermatozoa during/after ejaculation
(D) Any of these

36. Dag defect is an abnormlity of spermatozoa seen in its :
(A) Head (B) Neck (C) Tail (D) Acrosome cap

37. Dag defect is mainly due to :
(A) Reduced level of fructose in spermatozoa as well as seminal plasma
(B) Reduced level of zinc in spermatozoa as well as seminal plasma
(C) Due to an autosomal recessive gene
(D) Sign of disturbed spermatogenesis

38. In male the age of puberty can be well defined as a age at which :
(A) Partial separation of penis from prepucial frenulum occurs
(B) Animal shows sign of mounting
(C) Ejaculate contains sufficient spermatozoa to impregnate a cow
(D) All of these

39. What is the sequence of events with initiation of puberty in male :

i. Leydig cells secret testosterone
ii. Frequency of pulsatile release of LH increases
iii. Differentiation of sertoli cells
iv. Spermatogenesis

(A) i, ii, iii, iv (B) i, iii, ii, iv
(C) ii, iii, i, iv (D) ii, i, iii, iv

40. The duration of of one cycle of seminiferous epithelium in bull is :

(A) 4.5 days (B) 13.5 days
(C) 35.5 days (D) 61 days

41. For complete spermatogenesis .... days are needed.

(A) 4.5 days (B) 13.5 days (C) 35.5 days (D) 61 days

42. In bull the most effective stimuli for sexual stimulation is :

(A) Vision (B) Smell
(C) Voice (D) All are equally effective

43. Identification of estrus cow by licking and smelling of the external genitalia with curling of upper lip in a characteristic manner is called :

(A) Fergusson's reflex (B) Flehman's reflex
(C) Masturbation (D) Homosexuality

44. Match the correct one :

Species	**Gestation period**
(A) Cow	i. 280 days
(B) Sheep	ii. 150 days
(C) Buffalo	iii. 310 days
(D) Pig	iv. 113 days

(A) A-i, B-ii, C-iii, D-iv (B) A-iv, B-iii, C-ii, D-I,
(C) A-ii, B-i, C- iii, D-iv, (D) A-I, B-iv, C-iii, D-ii

45. Sustentacular cells that nurse the spermatozoa are also called :

(A) Sertoli cells (B) Leydig cells
(C) Sperm mother cells (D) Myoid cells

46. Which cells divide the seminiferous tubule in to basal and adluminal compartment ?

(A) Myoid cells (B) Sertoli cells
(C) Leydig cells (D) Spermatogonial cells

47. Erectile tissue or Cavernous tissue is comparable more than connective tissue of which species, which allows greater enlargement of penis during errection :

(A) Bull (B) Buffalo (C) Stallion (D) Ram

48. Primary meiotic division starts from :

(A) A sprmatogonia (B) B spermatogonia
(C) Primary Spermatocyte (D) Secondary Spermatocyte

49. The release of matured spermatocytes in to the lumen of seminiferous tubules is called :

(A) Spermatogenesis (B) Spermatocytogenesis
(C) Spermiation (D) Ejaculation

50. Implantation is invasive type in which species :
(A) Cow (B) Ewe (C) Doe (D) Human

51. Trans-uterine migration is a signal of maternal recognisation of pregnancy in :
(A) Ewe (B) Doe (C) Sow (D) Mare

52. Outer most membrane of placenta that is in contact of maternal endometrium is :
(A) Chorion (B) Allontois (C) Amnion (D) Yolk Sac

53. Which artery carries unoxygenated blood ?
(A) Umblical artery (B) Pulmonary artery
(C) Both of these (D) None of these

54. Placenta type in ruminants :
(A) Epithelia-chorial (B) Endothelio-chorial
(C) Hemo-chorial (D) Deciduate type

55. Match the following :

(a) Endothelio-chorial	i. Mare
(b) Diffuse type	ii. Ruminants
(c) Cotyledonary type	iii. Dog
(d) Hemo-chorial type	iv. Human

(A) a-i, b-ii, c-iii, d-iv (B) a-ii, b-i, c-iv, d-iii
(C) a-iii, b-i, c-ii, d-iv (D) a-i, b-iii, c-iv, d-ii

56. Among the following nutrients which one cannot cross placenta :
(A) Glucose (B) Fructose
(C) Electrolyte (D) Amino acid

57. PMSG is produced from :
(A) Syncytotrophoblastic cells (B) Endometrial cup
(C) Fetal tissue (D) Corpus luteum

58. Which hormone is the primary and ultimate necessity of onset of ovarian activity ?
(A) Estrogen (B) Progesterone
(C) FSH (D) LH

59. Among the following hypothalamic sites which one is not the surge center for LH release :
(A) Preoptic area (B) Arcuate nucleus
(C) Anterior hypothalamus (D) Suprachiasmatic nucleus

60. Hormone leptin is produced from :
(A) Skeletal muscle (B) Brown adipose tissue
(C) White adipose tissue (D) All of the above

61. Which among the following is a/an induced ovulatory ?
(A) Cat (B) Monkey (C) Camel (D) Rabbit

62. The quiescent stage of oocyte until puberty in pachytene stage of prophase 1 is called :
(A) Quiescent stage (B) Dormancy stage
(C) Dictyate stage (D) Pro-oocyte stage

63. The process of conversion of oogonia in to primary oocyte (mitosis) is called :
(A) Oogenesis (B) Folliculogenesis
(C) Ovogenesis (D) None of these

64. Which hormone have luteotropic action ?
(A) LH (B) PGF2α
(C) PRL (D) Both (A) and (C)

65. Luteolysin is carried from uterus to ovary through middle uterine vein except in ...., where it is via systemic circulation.
(A) Cow (B) Mare (C) Ewe (D) Sow

66. Primary site of luteolysis or luteolytic regression starts on :
(A) Large luteal cells (B) Small luteal cells
(C) Blood vessels (D) Thecal cells

67. .... is long day seasonal breeder.
(A) Mare (B) Ewe
(C) Buffalo (D) Sow

68. Nearly 80% of basal progesterone level during pregnancy is secreted by :
(A) Large luteal cells (B) Small luteal cells
(C) Non luteal cells (D) Thecal cells

69. Which type of luteal cells are enriched with LH receptors and can cause LH induced progesterone release ?
(A) Large luteal cells (B) Small luteal cells
(C) Both (A) and (B) (D) Thecal cells

70. Waxing in mare is associated with :
(A) Estrus (B) Pregnancy
(C) Placentation (D) Parturition

71. The uterine discharge that occurs normally in puerperium period is called :
(A) Involution (B) Lochia
(C) Uterine milk (D) Both (B) and (C)

72. The prostate is external in :
(A) Stallion (B) Dog
(C) Ram (D) Bull

73. Egg production drops in poultry during summer due to :
(A) Decreased humidity (B) Increased panting
(C) Decreased feed intake (D) Increased temperature

74. In bovines the right time to transfer embryo in recipient is :
(A) 6-7 days post estrus (B) On the day of estrus
(C) On 10th day of estrus (D) 3-4 days post estrus

75. Total time taken for spermatogenesis in bull is :
(A) 54 days (B) 42 days
(C) 84 days (D) 21 days

76. Hemochorial type of placentation found in primates, how many materno-fetal layers.
(A) 2 (B) 3 (C) 4 (D) 6

77. Synchronization of estrus by shortening the luteal process can be done by :

(A) Oral progesterone treatment

(B) Ear or vaginal implant of progesterone

(C) Estrogen and PGF2α injection

(D) Both (A) and (B)

78. When 2 PGF2α injections are given for synchronization of estrus then they are given at an interval of :

(A) 7 days (B) 9 days

(C) 11 days (D) 13 days

79. Following 2nd dose of PGF2α, after what interval AI should be done :

(A) Immediately (B) After 6-12 hours

(C) After 48-72 hours (D) With appearance of signs of AI

80. Among the following which one is not a synchronization protocol :

(A) Ovosync (B) Heatsync

(C) Cosync (D) Prosync

81. Superovulation is also called :

(A) Super-stimulation (B) Syncronisation

(C) Embryo transfer (D) Estrus induction

82. Embryos should be collected from recipient at what stage for embryo transfer :

(A) 2 cell stage (B) 4 cell stage

(C) Morula stage (D) Blastocyst stage

83. Blastocyst is collected from the recipient for embryo transfer on :

(A) 2 days post estrus (B) 5 days post estrus

(C) 7 days post estrus (D) 9 days post estrus

84. Trans-cervical method is not an advisable method of embryo transfer in sheep because :

(A) Short uterine horn (B) Tortuous oviduct

(C) Small size of cervical opening (D) All of the above

85. What is the best substance used as flushing medium during embryo collection ?

(A) RPMI (B) DPBS

(C) Citrate buffer (D) Both (B) and (C)

86. The flushing medium after embryo collection is allowed to pass through a filter of pore size :

(A) 10 μm (B) 60 μm (C) 150 μm (D) 300 μm

87. Vitrification is a method of cryopreservation in embryos, this method is also called :

(A) Conventional method of cryopreservation

(B) Non-conventional method of cryopreservation

(C) Equilibrium cryopreservation

(D) Both (B) and (C)

88. Thawing is done at :

(A) 0 degree centigrade (B) 10 degree centigrade

(C) 30 degree centigrade (D) 50 degree centigrade

89. Slow rate of cooling is associated with which problem :
(A) Solution effect (B) Mechanical damage
(C) Freezing shock (D) None of the above

90. Which one is a cryopreservating agent ?
(A) Sucrose (B) Ethylene glycol
(C) DMSO (D) All of the above

91. Which among the following is a not a method of embryo collection ?
(A) Slicing (B) Dissection
(C) Aspiration (D) All of the above

91. Which among the following is a not a method of embryo collection ?
(A) Slicing (B) Dissection
(C) Aspiration (D) All of the above

92. Which method of oocyte collection is considered best technique with respect to oocyte quality ?
(A) Slicing (B) Dissection
(C) Aspiration (D) Follicular puncture

93. Which method of oocyte collection is considered best technique with respect to rate of oocyte collection ?
(A) Slicing (B) Dissection
(C) Aspiration (D) Follicular puncture

94. For adequate oocyte maturation at least .... layers of cumulus is needed.
(A) Oocyte with atretic cumulus (B) Denuded oocyte
(C) 2 layers (D) 5 layers

95. Which one of the following is a marker of in-vitro oocyte maturation ?
(A) Oocyte resume meiosis and reaches metaphase II stage
(B) Expansion of cumulus
(C) Oocyte will extrude the 1st polar body
(D) All of the above

96. Which among the following is false for judging the soundness of boar ?
(A) Over 100 million sperm/ml should present
(B) More than 65% of motile sperm should exhibit forward progressive motility
(C) More than 80% of spermatozoa should of normal morphology
(D) All of the above

97. What is the volume bull semen per ejaculate ?
(A) 1-2 ml (B) 5-7 ml
(C) 30-50 ml (D) 150 ml

98. If sperms are swimming in a tight circular motion it may be due to :
(A) Cold shock (B) Aged sperm
(C) Stored sperm (D) Hypo-proteinemia

99. Sperm concentration can be measured by :
(A) Spectrophotometer (B) Hemocytometer
(C) Colorimeter (D) All of the above

100. Diadem defect is seen in :
(A) Head (B) Middle piece
(C) Principal pica (D) Tail

101. Voluminous semen is feature of :
(A) Cattle and sheep (B) Swine and Horse
(C) Sheep and Goat (D) Swine and Bird

102. Rheological property of cervical mucus include :
(A) Thixotrophy (B) Tack
(C) Spinbarkeit (D) All of the above

103. Uptake of sperm by phagocytes in female reproductive tract is called :
(A) Spermateliosis (B) Spermophagy
(C) Spermiation (D) Leuko-uptake

104. As in horse and swine sperm reservoirs are located in utero-tubal junction, in dog the reservoir is at :
(A) Utero-tubal junction (B) Uterus
(C) Oviduct (D) Endometrial cups

105. What is the pH of vagina and Cervix respectively in cow ?
(A) 4.0 and 8.4 (B) 8.4 and 4.0
(C) 7.1 and 4.0 (D) 4.0 and 7.1

106. Sperm loss in female reproductive tract occurs by which of the following way ?
(A) Vaginal back flow (B) Loss in to peritoneal cavity
(C) Spermophagy (D) All of the above

107. In most species fertile life of egg is :
(A) 12-24 hrs. (B) 72 hrs.
(C) 98 hrs. (D) 7 days

108. The fertilizable life of egg in mare and sow is :
(A) 8- 10 hrs. (B) 12-24 hrs.
(C) 72 hrs. (D) 98 hrs.

109. Sperm head has DNA coupled with arginine rich protein, this DNA protein complex is called :
(A) Histone (B) Protamine
(C) Argimine (D) Nucleomine

110. Acrosome contains all the following except :
(A) Hyaluronidic acid (B) Hyaluronidase
(C) Acrosin (D) Aryl sulphate

111. Which segment of acrosome 1st fuses with oocyte membrane during fertilization ?
(A) Basal segment (B) Equatorial segment
(C) Annular membrane (D) None of the above

112. Annulus is present between :
(A) Acrosome and head (B) Head and middle piece
(C) Middle piece and principal piece (D) Principal piece and end piece

113. Residual cytoplasm when present in neck region it is called :
(A) Annulus (B) Residual cytoplasm
(C) Proximal cytoplasm (D) Distal cytoplasm

114. Acrosin of acrosome helps in digesting :
(A) Zona pelucida (B) Cumulus
(C) Plasma membrane (D) Theca layer

115. The metamorphic changes that occur in spermatid via progressive structural and developmental alternation is called :
(A) Spermatocytogenesis (B) Spermiogenesis
(C) Spermiation (D) All of the above

116. Hyaluronidase of acrosome helps in digesting :
(A) Zona pelucida (B) Cumulus
(C) Plasma membrane (D) Theca layer

117. Blood testis barrier is mainly made by what types of cells :
(A) Myoid cells and Sertoli cells
(B) Myoid cells and Leydig cells
(C) Sertoli cells and Leydig cells
(D) Spermatogonial cells and Sertoli cells

118. Principal permeability barrier of blood testis barrier is made up off :
(A) Myoid cells (B) Sertoli cells
(C) Leydig cells (D) Spermatogonial cells

119. Testosterone produced by leydig cells is carried by the help of ....
(A) Inhibin (B) Androgen binding protein
(C) Estrogen (D) Activin

120. Sertoli cells secrets FSH stimulant Activin, which is similar to :
(A) α unit of Inhibin (B) Dimer of α unit of Inhibin
(C) β unit of Inhibin (D) Dimer of β unit of Inhibin

121. Transport of spermatozoa through epididymis in bull takes :
(A) 2 days (B) 5 days (C) 7 days (D) 12 days

122. Unnecessary metabolism of spermatozoa in epididymis is prevented in lab animals and bull is .... and .... respectively.
(A) Immobilin and quiescence factor respectively
(B) quiescence factor and immobilin respectively
(C) ABP and quiescence factor respectively
(D) quiescence factor and ABP respectively

123. In most species during ovulation ovum is in metaphase II of 2nd meiotic division except in ....
(A) Goat and dog (B) Sheep and goat
(C) Dog and horse (D) Horse and swine

124. Which among the following is false about capacitation ?
(A) Destabilization of plasma membrane occur by efflux of cholesterol and influx of calcium ion
(B) Membrane permeability changes

(C) Decrease in protein tyrosine kinase activity results
(D) Major site of capacitation is isthmus

125. Sperm longevity in female reproductive tract is highest in :
(A) Cow (B) Goat
(C) Horse (D) Dog

126. Axoneme development starts in which stage of spermiogenesis :
(A) Golgi phase (B) Cap phase
(C) Acrosomal phase (D) Maturation phase

127. Longevity of bull sperm in female reproductive tract is :
(A) 8 hrs. (B) 12 hrs.
(C) 48 hrs. (D) 72 hrs.

128. .... is also called zonalysin.
(A) Hyluronidase (B) Acrosin
(C) Aryl sulphate (D) All of the above

129. Which among the following is a structural protein receptor of zona pelucida ?
(A) ZP1 (B) ZP2
(C) ZP3 (D) Both (A) and (B)

130. .... is the identification that acrosomal reaction has started.
(A) Vesiculation (B) Engulfment
(C) Polyspermy (D) Both (A) and (B)

131. Zona block is important for :
(A) Completion of Acrosomal Reaction
(B) Prevention of Polyspermy
(C) Successful engulfment
(D) All of the above

132. What is the site of polar body extrusion in ovum ?
(A) Animal pole (B) Vegetal pole
(C) Vegetative pole (D) Annual pole

133. Cleavage divisions are :
(A) Only Mitotic (B) Only Meiotic
(C) Initially Mitotic then Meiotic (D) Initially Meiotic then Mitotic

134. After .... stage of zygote synchrony in division of blastomeres are lost.
(A) Two cell stage (B) 4 cell stage
(C) Morula stage (D) Blastula stage

135. The combined process of blastomere flattening and asymmetric distribution of microvilli is called :
(A) Vesiculation (B) Synchrony
(C) Cleavage (D) Compaction

136. In the blastocyst stage blastomeres are present at one portion of ectoderm known as :
(A) Embryoblast (B) Inner cell mass
(C) Trophoblast (D) Both (A) and (B)

137. Zona hatching occurs at .... days post ovulation.

(A) Immediately after ovulation (B) 4 days

(C) 7 days (D) 9 days

138. In polytoccous species what among the following is important for embryo survival :

(A) Intrauterine migration

(B) Equidistance spacing between embryos

(C) Change in morphology of blastomere

(D) Both (A) and (B)

139. In which species trans-uterine migration is common :

(A) Cattle (B) Human (C) Sheep (D) Horse

140. In which species intrauterine migration is common :

(A) Cattle (B) Human (C) Sheep (D) Horse

141. In pig at least .... no. of embryos per uterine horn is needed for successful pregnancy.

(A) 1 (B) 2 (C) 3 (D) 4

142. Invasive type of implantation seen in :

(A) Cattle (B) Human (C) Sheep (D) Horse

143. Maternal recognition of pregnancy in cattle occurs at :

(A) 5th day of pregnancy (B) 9th day of pregnancy

(C) 12th day of pregnancy (D) 16th day of pregnancy

144. Maternal recognition of pregnancy is related to :

(A) INF ζ (B) INF γ

(C) INF α (D) Both (A) and (B)

145. Which among the following is true about implantation ?

(A) Elongation of trophoectoderm is essential for INF ζ production

(B) Muc- 1 is a adhesion factor that is abundant in receptive phase of pregnancy

(C) Muc-1 protein abundance is independent of progesterone

(D) All the above are true

146. Trophoblastic cells of conceptus migrate and fuse with uterine luminal epithelial cells to form :

(A) Caruncules (B) Carbuncules

(C) Conceptus (D) Syncytium

147. Which among the following are multiple adhesion proteins expressed during adhesion process of implantation ?

(A) Muc-1 and INF-ζ (B) GlyCAM-1 and Galectin-15

(C) Integrin and UTMP (D) Both (B) and (C)

148. UTMP stands for :

(A) Uterine trophoblastic membrane protein

(B) Uterine milk protein

(C) Unassociated trophoblastic milk protein

(D) Unusual trophoblastic membrane protein

149. Which among the following is true about INF-ζ ?
(A) It is produced by trophoblastic cells
(B) It is a low molecular weight acidic protein
(C) It regulates production of PGF2α
(D) All the above are true

150. During pregnancy under the influence of progesterone uterine environment gets insensitive to estrogen and oxytocin it is called :
(A) Maternal recognition of pregnancy (MRP)
(B) Fergussion's reflex
(C) Progesterone block
(D) Both (A) and (C)

151. Endocrine- exocrine model of MRP is seen in :
(A) Cow (B) Pig
(C) Horse (D) Dog

152. Endometrial glands open in to the endometrial lumen except in :
(A) Caruncular area (B) Middle part of uterus
(C) Utero-tubal junctional area (D) Utero-cervical junction

153. Fetus is similar to :
(A) Isograft (B) Allograft
(C) Normograft (D) None of the above

154. Heterospermic insemination refers to :
(A) Use of semen from one male to inseminate more no. of female
(B) Use of semen from more than one male in the insemination dose
(C) Use of semen from different species to prepare one insemination dose
(D) All are true

155. Pregnancy occurs most frequently cow :
(A) Right horn (B) Left horn
(C) Equal chance in each horn (D) Cannot be said

156. Normal Litter size in pig is :
(A) 6 (B) 8 (C) 10 (D) 12

157. Internal temperature of Artificial vagina in cattle is :
(A) 30 degree (B) 37 degree
(C) 43 degree (D) 47 degree

158. Electroejaculation technique of semen collection is mainly used in ram and boar with voltage varies with :
(A) 2-20 V (B) 3-30 V (C) 4-40 V (D) 5-50 V

159. Normal pH of bovine semen is :
(A) 5.6 (B) 6.8 (C) 7.4 (D) 8.4

160. Semen collection is of longer duration in which species :
(A) Cattle (B) Buffalo (C) Sheep (D) Swine

161. Biological activity of FSH, for FSH specific activity is located in :
(A) Subunit A (B) Subunit B
(C) Both (A) & (B) (D) None of these

162. Egg production drops in poultry due to :
(A) Decreased humidity (B) Increases panting
(C) Decreased feed intake (D) Increased temperature

163. Artificial insemination in cattle was done for the first time in India at :
(A) Chennai (B) Mysore
(C) Mumbai (D) Agra

164. Chromosome number in cattle is :
(A) 60 (B) 58 (C) 62 (D) 48

165. Oestrous length in sow is :
(A) 21 days (B) 24 days
(C) 18 days (D) 15 days

166. Act of parturition in rabbit is called :
(A) Whelping (B) Kindling
(C) Kidding (D) Calving

167. Diluent used for production of chilled bull semen :
(A) Tris (B) Egg Yolk citrate
(C) Both (A) and (B) (D) None of these

168. Species in which sigmoid flexure is post-scrotal :
(A) Cattle (B) Sheep
(C) Goat (D) All of the above

169. Diluent used in frozen bull semen production is called :
(A) EYC (B) Tris
(C) Skim milk (D) None of these

170. Which one is correctly matched (Species with act of copulation) ?
(A) Cattle and buffalo- serving (B) Pig- Coupling
(C) Dog- Tupping (D) All of the above

171. Species in which seminal plasma should be removed from semen before freezing :
(A) Cattle (B) Goat
(C) Buffalo (D) None of these

172. The left ovary is more functional than right ovary :
(A) Sow (B) Ewe
(C) Doe (D) Cow

173. The ingredient in semen diluent which is a cryoprotectant :
(A) Glycerol (B) Tris
(C) Phosholipids (D) Lecithine

174. Right ovary is functional in :
(A) Cow (B) Pig
(C) Mare (D) Fowl

175. The preovulatory surge of Luteinizing hormone is initiated by an increase in the circulating level of the following hormone :
(A) Estrogen (B) Progesterone
(C) Prolactin (D) Cortisol

176. Which hormone is responsible for maternal behaviour ?
(A) Estrogen (B) Oxytocin
(C) Prolactin (D) Progesterone

177. Inhibin is also known as :
(A) Folligon (B) Folliculosomatostatin
(C) Folliculostatin (D) Follicle inhibiting hormone I

178. Inhibin is secreted from which cell :
(A) Epididymis (B) Leydig cells
(C) Sertoli cells (D) All of the above

179. The secretion of melatonin is increased in sheep, during the exposure of :
(A) Light period (B) Dark period
(C) None of these (D) Both (A) & (B)

180. The sign of estrus in goats :
(A) Frequent bleating (B) Frequent wagging of tail
(C) Edematous and reddened vulva (D) All of the above

181. Species in which female are heterogametic :
(A) Avian (B) Bovine (C) Caprine (D) Canine

182. The largest segment of avian oviduct :
(A) Infundibulum (B) Magnum (C) Isthmus (D) Uterus

183. Part of poultry oviduct where fertilization takes place :
(A) Infundibulum (B) Isthmus (C) Magnum (D) Uterus

184. Part of poultry oviduct which secrets major portions of albumen :
(A) Infundibulum (B) Isthmus
(C) Magnum (D) Uterus

185. Part of poultry oviduct which secrets calcareous shell :
(A) Infundibulum (B) Isthmus
(C) Magnum (D) Uterus

186. The shell gland of oviduct of poultry is :
(A) Infundibulum (B) Isthmus
(C) Magnum (D) Uterus

187. Hormone responsible for broodiness of poultry is :
(A) Estrogen (B) Progesterone
(C) Prolactin (D) Oxytocin

188. An extender used for room temperature preservation of bull semen :
(A) Egg yolk citrate diluent
(B) Coconut milk Extender
(C) Egg yolk sodium bicarbonate Glucose milk diluent
(D) Tris diluent

189. Per rectal examination is done in cattle :
(A) To palpate the accessible organs in the pelvic and abdominal cavities
(B) To detect estrus and perform AI
(C) For pregnancy Diagnosis
(D) All of the above

190. Average number of spermatozoa per milliliter of bull semen :

(A) 800-1200 millions (B) 80-100 millions
(C) 3000 millions (D) 30 millions

191. Ideal number of inseminations required per conception in cattle :

(A) 1 (B) 1.6 (C) 2.5 (D) 3

192. Castration is done in animals to :

(A) Fatten (B) Make docile
(C) Avoid breeding (D) All of the above

193. Castration in animals can be done by :

(A) Surgical removal (B) Crushing spermatic cord
(C) Removing part of epididymis (D) All of the above

194. Flushing of sows and gilts before breeding :

(A) Increases ovulation rate (B) Increases libido
(C) Reduces breeding stress (D) None of the above

195. pH of bull semen :

(A) 5.8 (B) 6.8 (C) 7.8 (D) 8.8

196. Polyspermy during fertilization of mammalian ovum is not usually possible :

(A) Because sperms digest one another by their own enzymes
(B) Because other sperms lose their tails and cannot penetrate
(C) Because the acrosomal enzyme of the penetrating sperm diffuses in to mucopolysaccharide lattice of the zona-pellucida
(D) All of the above

197. Capacitation of ejaculated sperm can be possible :

(A) In the female genitalia
(B) In the artificial medium
(C) Both (A) and (B)
(D) In the female genitalia, In the artificial medium and gastro intestinal tract also

198. Hormone Relaxin is secreted from :

(A) Corpus Luteum of the pregnancy
(B) Corpus Luteum and endometrium of the non-pregnant uterus during secretory phase
(C) Syncytiotrophoblast of placenta
(D) All of the above

199. The female animal in which the deposition of semen takes place inside uterus as the penis of the passes through the cervix during copulation in :

(A) Swine (B) She-buffalo
(C) Ewe (D) Bitch

200. Secretion of seminal vesicle :

(A) Is alkaline (B) Is of no importance in reproduction
(C) Is mucoid (D) None of these

201. Boar seminal vesicle contains lots of :

(A) Fructose (B) Citric acid
(C) Ergothionine (D) None of these

202. The primary role of pampiniform plexus is to regulate :
(A) Temperature of testis (B) Blood flow to testis
(C) Nerve supply to testis (D) None of these

203. P.M.S.G has :
(A) F.S.H. like action (B) L.H like action
(C) Both (A) and (B) (D) PRL like action

204. The placenta of a bitch is :
(A) Diffuse type (B) Syndesmochorial
(C) Endotheliochorial (D) All of the above

205. Which animal has shortest pregnancy period ?
(A) Sheep (B) Goat
(C) Bitch (D) Horse

206. When a pig in heat hears the sound of a boar then it :
(A) Runs away (B) Runs towards boar
(C) Stand rigid without moving (D) None of these

207. Sumer sterility in rams is due to :
(A) Increase in thyroxine levels
(B) Decrease in thyroxine level
(C) Normal thyroxine but low level of FSH and LH
(D) Normal thyroxine but high level of FSH and LH

208. The urine of which animal is richest source of estrogen :
(A) Stallion (B) Buck
(C) Pregnant cow (D) Pregnant ewe

209. The animal in which pheromone is secreted in saliva that can attract female in heat and cause immobilization :
(A) Mare (B) Stallion
(C) Boar (D) Bull

210. Signs of frequent estrus in irregular interval is called :
(A) Nymphomaniac (B) Buller-steer syndrome
(C) Split estrus (D) Homosexuality

211. Folligon is a commercial name of which hormone :
(A) eCG (B) FSH
(C) LH (D) Progestron

212. .... hormone controls the oviposition.
(A) Estrogen (B) Progesterone
(C) Oxytocin (D) Prolactin

213. The body temperature of cattle is :
(A) 100-102°F (B) 104°F
(C) 99°F (D) 105-107°F

214. The body temperature of fowl is :
(A) 100-102°F (B) 104°F
(C) 99°F (D) 105-1070F

215. The term flushing means :
(A) To clean the animal using sprayer
(B) Feeding liberally before breeding of sheep
(C) Washing the sheep before shearing
(D) Vaccination of sheep before breeding season

216. Feeding extra concentrate to pregnant sheep is known as :
(A) Steaming-up (B) Flushing
(C) Holding (D) Heeding

217. Gestation period of camel is :
(A) 280 days (B) 310 days
(C) 340 days (D) 390 days

218. Gestation period of horse is :
(A) 280 days (B) 310 days
(C) 340 days (D) 390 days

219. Removal of ovary is called as :
(A) Spaying (B) Castration
(C) Docking (D) None of these

220. Reduction of organ to its original size is called :
(A) Innovation (B) Invocation
(C) Involution (D) Evolution

— Answer Sheet —

Q. No.	Ans	Q. No.	Ans	Q. No.	Ans	Q. No.	Ans	Q. No.	Ans	Q. No.	Ans
1.	B	21.	C	41.	D	61.	B	81.	A	101.	B
2.	A	22.	C	42.	A	62.	C	82.	D	102.	D
3.	C	23.	C	43.	B	63.	C	83.	B	103.	B
4.	B	24.	C	44.	A	64.	D	84.	B	104.	D
5.	A	25.	D	45.	A	65.	B	85.	B	105.	A
6.	D	26.	C	46.	B	66.	A	86.	B	106.	D
7.	D	27.	B	47.	C	67.	A	87.	B	107.	A
8.	D	28.	C	48.	C	68.	A	88.	C	108.	A
9.	C	29.	D	49.	C	69.	B	89.	A	109.	B
10.	D	30.	D	50.	D	70.	D	90.	D	110.	A
11.	B	31.	D	51.	D	71.	B	91.	D	111.	B
12.	B	32.	D	52.	A	72.	B	92.	A	112.	C
13.	A	33.	B	53.	C	73.	B	93.	C	113.	C
14.	C	34.	B	54.	A	74.	A	94.	C	114.	A
15.	A	35.	C	55.	C	75.	A	95.	D	115.	B
16.	C	36.	C	56.	B	76.	B	96.	D	116.	B
17.	C	37.	B	57.	B	77.	C	97.	B	117.	A
18.	C	38.	C	58.	D	78.	C	98.	A	118.	B
19.	A	39.	D	59.	B	79.	C	99.	D	119.	B
20.	B	40.	B	60.	C	80.	D	100.	A	120.	D

Q. No.	Ans	Q. No.	Ans	Q. No.	Ans	Q. No.	Ans	Q. No.	Ans	Q. No.	Ans
121.	C	138.	D	155.	A	172.	A	189.	D	206.	C
122.	A	139.	C	156.	C	173.	A	190.	A	207.	B
123.	C	140.	D	157.	C	174.	A	191.	B	208.	A
124.	C	141.	B	158.	B	175.	A	192.	D	209.	C
125.	C	142.	B	159.	B	176.	C	193.	D	210.	A
126.	B	143.	D	160.	D	177.	C	194.	A	211.	A
127.	C	144.	A	161.	B	178.	C	195.	B	212.	C
128.	B	145.	A	162.	B	179.	B	196.	C	213.	A
129.	D	146.	D	163.	B	180.	D	197.	C	214.	D
130.	A	147.	D	164.	A	181.	A	198.	D	215.	B
131.	B	148.	B	165.	D	182.	B	199.	A	216.	A
132.	A	149.	D	166.	B	183.	A	200.	A	217.	D
133.	A	150.	C	167.	B	184.	C	201.	C	218.	C
134.	C	151.	B	168.	D	185.	D	202.	A	219.	A
135.	D	152.	A	169.	B	186.	D	203.	A	220.	C
136.	D	153.	B	170.	D	187.	C	204.	C		
137.	D	154.	B	171.	B	188.	B	205.	C		

• Chapter 9 •

Sense Organ

1. **Ear pinna is mobile in :**
(A) Cattle (B) Horse
(C) Pig (D) Both (A) and (B)

2. **Middle ear has 3 small bones (outer to inner) named as :**
(A) Mallus-Incus-Stapes (B) Incus-Mallus-Stapes
(C) Stapes-Incus-Mallus (D) Stapes-Mallus-Incus

3. **The outer most part of the ear made of cartilage that connects to the auditory canals is :**
(A) Ear drum (B) Cochlea (C) Pinna (D) Oval window

4. **In case of birds the 3 bones are absent but a bony structure is present called as :**
(A) Ossicles (B) Helicotrema
(C) Collumela (D) Sclera tympany

5. **.............. connects from middle ear to brain.**
(A) Auditory tube (B) Semicircular canal
(C) Eustachian tube (D) Cochlea

6. **Fluid filled tubes attached to the cochlea and nerves in the inner ear that send messages about balance and head position to the brain (helps in body balance) :**
(A) Auditory tube (B) Semicircular canal
(C) Eustachian tube (D) Cochlea

7. **Spiral shaped part that transforms sound into nerve impulses and send it to brain is (helps in hearing) :**
(A) Auditory canal (B) Semicircular canal
(C) Eustachian tube (D) Cochlea

8. **Outer bony labyrinth is filled with fluid called :**
(A) Aqueous humor (B) Perilymph
(C) Endolymph (D) Auditory lymph

9. **Outer bony labyrinth is filled with fluid called :**
(A) Aqueous humor (B) Perilymph
(C) Endolymph (D) Auditory lymph

10. **The receptor cells that can sense sound are located in :**
(A) Tympanic membrane (B) Vestibular membrane
(C) Basilar membrane (D) Mucous membrane

11. **Some sounds can enter the inner ear through the bones of skull. This is called :**
(A) Bony adaptation (B) Bony conduction
(C) Bony reception (D) Bony transmission

12. **Which cranial nerve carries the neural information regarding sound to brain :**
(A) 3rd (B) 5th (C) 8th (D) 9th

13. **How many semicircular canals are there in inner ear :**
(A) 2 (B) 3 (C) 4 (D) 5

14. Semicircular canals are filled with :
(A) Aqueous humor (B) Perilymph
(C) Endolymph (D) Auditory lymph

15. Orderly arrangement between frequency of sound wave and region of cochlea (particularly basilar membrane) is called :
(A) Tonotopic organization (B) Young.Helmholtz theory
(C) Helicotrema (D) Both (A) and (B)

16. The waxy material in the external auditory canal (meatus) is the product of :
(A) Tarsal gland (B) Ceruminous gland
(C) Lacrymal gland (D) Bowman's gland

17. Apart from 3 ossicles, middle ear also have 2 small skeletal muscles namely :
(A) Malledius and Stapedius (B) Stapedius and tensor tympany
(C) Malledius and tensor tympany (D) Stapedius and insipidus

18. Bony labyrinth of vestibular system contain :
(A) 3 semicircular duct (B) Otolith organ
(C) Both (A) and (B) (D) Organ of Corti

19. The transmission of sound waves through the tympanic cavity of the middle ear occurs through :
(A) Air (B) Membrane (C) Bone (D) fluid

20. .... separate middle ear from inner ear.
(A) Tensor tympany (B) Basilar membrane
(C) Tympanic membrane (D) Oval window

21. We can hear the sound frequency range from :
(A) 20 to 20,000 Hz (B) 28 to 20,000 Hz
(C) 52 to 20,000 Hz (D) 100 to 20,000 Hz

22. Guttural pouch of horse is a large diverticulum seen in :
(A) Eustachian tube (B) Semicircular canal
(C) Otolith organ (D) Ampulla

23. .... nucleus of brain receives the auditory information at last.
(A) Lateral geniculate nucleus (B) Medial geniculate nucleus
(C) Ventro-medial nucleus (D) Both (A) and (B)

24. Time taken for transmission of sound to brain is called latent period, which is about :
(A) 2 milli second (B) 5 milli second
(C) 10 milli second (D) 30 milli second

25. Both eyes seeing the same object is called :
(A) Long sightedness (B) Unitary vision
(C) Binocular vision (D) Ocular vision

26. Which of the following is not part of the vascular layer ?
(A) Cornea (B) Choroid (C) Ciliary body (D) Iris

27. Which region of eye can sense light and can transform it to nerve impulse ?
(A) Iris (B) Retina (C) Cornea (D) Pupil

28. Clear watery fluid that circulates in the front part of the eye and keeps constant pressure within the eye :

(A) Aqueous humor (B) Vitreous humor
(C) Tear (D) Cornea

29. The circular opening in the colored part of the eye that opens and contracts to let in more or less light :

(A) Iris (B) Pupil (C) Choroid (D) Retina

30. This part focuses light, changing shape as it takes in reflected light from objects near and far :

(A) Iris (B) Pupil (C) Lens (D) Ciliary muscle

31. Which one is not a sign of Horner's syndrome ?

(A) Enopthalmia (B) Protrusion of eyelid/Blepheritis
(C) Myopic vision (D) Miosis

32. Accommodation is mainly provided by :

(A) Cornea (B) Lens (C) Pupil (D) Iris

33. Retina contains .... number of different cells.

(A) 3 (B) 4 (C) 5 (D) 6

34. .... is appoint on retina responsible for accurate vision.

(A) Cones (B) Rods (C) Fovea (D) Tapetum

35. Photopigment in cones is :

(A) Inopsin (B) Idopsin (C) Rhodopsin (D) Conopsin

36. Rods are important for .... vision.

(A) Day vision (B) Night vision
(C) Colour vision (D) Both (B) and (C)

37. A light reflecting medium prominent in nocturnal animal present between photoreceptor cell layer of retina and choroid is called :

(A) Fovea (B) Helicotrema
(C) Tapetum (D) Amacrine cell

38. In birds fovea contains only .... cells.

(A) Cones (B) Rods
(C) Amacrine cells (D) Ganglionic cells

39. What is the correct sequence of event that occur when light fall on photoreceptor cells :

1. Retinal component undergo transformation (11 cis retenene to 11 trans retenene)
2. Activation of opsin
3. Hyperpolarization
4. Increase in intracellular second messenger

(A) 1-2-3-4 (B) 2-1-3-4 (C) 1-2-4-3 (D) 2-1-4-3

40. Particular population of cones are assigned by brain to recognize a colour of particular frequency is called :

(A) Tonotopic organization (B) Visual pathway
(C) Young-Helmholtz theory (D) None of these

41. Dim light vision is also called :
(A) Scotopic vision (B) Photopic vision
(C) Nocturnal vision (D) Dicoptic vision

42. Which one of these is not an exteroceptor ?
(A) Photoreceptor of eye (B) Chemoreceptor of tongue
(C) Proprioceptor (D) Tactile receptor

43. Which of the following senses involve chemoreceptor ?
(A) Smell and taste (B) Hearing and vision
(C) Smell and vision (D) Smell, taste and vision

44. Olfactory neurons project from olfactory bulb to .... of cerebral cortex.
(A) Frontal lobe (B) Temporal lobe
(C) Parietal lobe (D) Occipital lobe

45. Taste buds are located on :
(A) Fungiform papillae (B) Circumvallate papillae
(C) Filliform papillae (D) Both (A) and (B)

46. Ruminant calves exhibit preference for :
(A) Glucose (B) Salt (C) Sucrose (D) Acetic acid

47. Which of the following is an example of microsmatic animal ?
(A) Pig (B) Dog (C) Cow (D) Sea whale

48. Olfactory glands are also called :
(A) Nasal gland (B) Bowman's gland
(C) Salt gland (D) Mucus gland

49. Taste receptors are called :
(A) Taste buds (B) Olfactory receptor
(C) Gustatory receptor (D) Taste pore

50. Which of the following organs does not contain hair cells as sensory receptor ?
(A) Organ of Corti (B) Nose
(C) Saccules (D) Taste buds

51. Which one of the following explains adaptation of receptor :
(A) Adaptation appears with a low intense sensation produced by steady stimulus applied over time
(B) A high intense stimulus with repeated stimuli resulted into adaptation
(C) Adaptation appears early when other stimuli are applied simultaneously
(D) All are true

52. Which of the following sensory receptor adapts rapidly ?
(A) Nocireceptor (B) Thermo receptor
(C) Propioreceptor (D) Chemoreceptor

53. Propioceptors convey information from :
(A) Muscle (B) Joint
(C) Semicircular canals of inner ear (D) All of the above

54. Two point touch threshold test serve to demonstrate :
(A) Lateral inhibition (B) Sensory adaptation
(C) Receptive field (D) All of the above

55. The senses of smell and taste have all of the following in common except that they :

(A) are both based on exteroceptors
(B) can only sense molecules dissolved in liquid.
(C) both transmit to the brain via the glossopharyngeal nerve (IX)
(D) are both based on chemoreceptors.

56. Angular acceleration of body is primarily detected by :

(A) Utricle (B) Semicircular canal
(C) Organ of Corti (D) Saccule

57. The hair cells of semicircular canal located on the :

(A) Otolith organ (B) Ampulla
(C) Scala media (D) Otolith membrane

58. Which of the following does not belong to the rest ?

(A) Amacrine cell (B) Ganglionic cell
(C) Rod cell (D) Horizontal cell

59. The light intensity absorbed best by rhodopsin is :

(A) Violet (B) Red (C) Green (D) Yellow

60. Different groups of cone cells are specialized for absorption in all the following regions of the visible spectrum except :

(A) Violet (B) Red (C) Green (D) Yellow

61. The optic nerve is composed of one axon fiber extending from each .... cell in the retina.

(A) Ganglionic cell (B) Amacıine cell
(C) Bipolar cell (D) Horizontal cell

62. Cylindrical lenses are used to correct :

(A) Myopia (B) Hyperopia (C) Presbyopia (D) Astigmatism

63. What sensory receptors detect pheromones ?

(A) Gustatory (B) Auditory (C) Olfactory (D) Photoreceptor

64. Gustatory receptors are linked with :

(A) Smell (B) Taste (C) Hearing (D) Both (A) and (B)

65. In man how many primary taste sensation are there :

(A) 2 (B) 3 (C) 4 (D) 5

66. .... maintains the pressure of eyeball.

(A) Iris (B) Cornea
(C) Aqueous humor (D) Pupillary sphincter muscle

67. Reduction in pupillary size is called :

(A) Meiosis (B) Miosis
(C) Mydriasis (D) Accommodation

68. The phenomenon of change in the curvature of lens to see the far and near object is called :

(A) Miosis (B) Mydriasis
(C) Accomodation (D) Binocular vision

69. Which among the following are retinal cells :

(A) Amacrine and Horizontal cells
(B) Ganglionic and Photoreceptor cells
(C) Amacrine and Photoreceptor cells
(D) All the above type of cells

70. Majority of photoreceptor cells in nocturnal animals are :

(A) Rods (B) Cones (C) Both equal (D) Not fixed

71. The photo chemical in rods is rhodopsin likewise in cones it is :

(A) Rhodopsin (B) Chonopsin (C) Inopsin (D) Iodopsin

72. When photon strikes the photo receptor cell :

(A) 11 trans retinene is converted to all trans retinene
(B) 11 trans retinal is converted to all trans retinol
(C) 11 cis retinal is converted to all trans retinal
(D) 11 cis retinol is converted to all trans retinol

73. A small area of retina having no photoreceptor cell is known as :

(A) Fovea (B) Area centralis
(C) Blind spot (D) Optic chiasma

74. The Tapetum Lucidum is present in :

(A) Sclera (B) Choroid (C) Iris (D) Retina

75. Hyperpolarization of Rod cells is caused by :

(A) Closure of cGMP- gated Na^+ channels
(B) Opening of cGMP- gated Na^+ channels
(C) Increased activity of Na^+ K^+ pumps
(D) Closure of non-gated K^+ channels

76. The receptor that detects angular acceleration and deceleration of the head is :

(A) Organ of Corti (B) Macula
(C) Crista ampullaris (D) Saccule

77. Loss of ability to smell is called :

(A) Anosmia (B) Asomnia
(C) Insomnia (D) Hyposmia

78. The lifespan of taste receptor cells are about :

(A) 1 day (B) 10 days
(C) 30 days (D) 60 days

79. The orderly physical arrangement of basilar membrane of ear to respond different frequency of sound is called :

(A) Tonotopy (B) Auditory reflex
(C) Acoustic response (D) All of the above

80. Painful sensations are sensed by .... receptors.

(A) Propioreceptors (B) Nocireceptor
(C) Mechanoreceptors (D) Radioreceptors

81. According to Young Helmholtz theory there are .... types of cones.

(A) 2 (B) 3
(C) 4 (D) 5

82. When photon excites the rod cells the following is the sequence of event occur in rhodopsin :
(A) Rhodopsin- Lumirhodopsin- Metarhodopsin- Scotopsin + Retinene
(B) Rhodopsin- Metarhodopsin- Scotopsin- Lumirhodopsin + Retinene
(C) Rhodopsin- Metarhodopsin- Scotopsin- Lumirhodopsin + Retinene
(D) Rhodopsin- Metarhodopsin- Lumirhodopsin- Scotopsin – Retinene

83. Vision with both the eye at a time is called :
(A) Peripheral vision (B) Monocular vision
(C) Binocular vision (D) Dichromatic vision

84. What do you mean by multimodal receptor ?
(A) They respond to more than one type of stimuli
(B) They respond to same type of stimuli of different intensity
(C) Both (A) and (B)
(D) They are present in different site but respond to same type of stimuli

85. What do you mean by sensory adaptation ?
(A) Change of receptor type of an area due to a constant stimuli
(B) Ability to respond a stimuli increases with constant and prolong exposure to a stimuli
(C) Ability to respond a stimuli decreases with constant and prolong exposure to a stimuli
(D) Both (A) and (B)

86. Slowly adapting receptors are :
(A) Mechanoreceptor (B) Noci receptor
(C) Propio receptor (D) Thermoreceptor

87. Signs of Horner's syndrome :
(A) Miosis(small pupil diameter)
(B) Enopthalmia (eyeball move in to the socket)
(C) Protrusion of eyelid
(D) All are true

88. Alpha group of fibers have :
(A) Nearly 20μm diameter and are mylinated
(B) Conduction speed is highest
(C) Highest axon diameter among the primary afferent nerve fibers
(D) All are true

89. 2nd order neuron can be a :
(A) Interneuron (B) Ascending neuron
(C) Descending neuron (D) All of these

90. There are how many ascending pathways :
(A) 1 (B) 2 (C) 3 (D) 4

91. Due to damage to tissue previously non-painful stimuli produces pain is called :
(A) Allodynia (B) Hyperelgia (C) Hyposmia (D) None of these

92. Previously painful stimuli produces severe pain of greater magnitude and duration is called :
(A) Allodynia (B) Hyperelgia (C) Hyposmia (D) None of these

93. Which one is true about central hyperelgia :

(A) Occurs in dorsal horn of spiral cord

(B) It is also called primary hyperalgia

(C) Occurs when endogenous chemicals released at the site of tissue injury

(D) All of these are true

94. Which one is true about stretch reflex :

(A) It is a spinal reflex also called as myotatic reflex

(B) It is a monosynaptic reflex

(C) Sensory receptors are at muscle spindle

(D) All of these true

95. Which among is true about muscle spindle :

(A) These are widely scattered in skeletal muscle belly

(B) They have fusiform/spindle like appearance

(C) Muscle fibers with in the spindle are called intrafusal fibers and outside the spindle are called extrafusal fibers

(D) All of these are true

96. Which among the following is false about Golgi tendon reflex :

(A) It is a monosynaptic reflex

(B) Prevents tendon tearing

(C) Sensory receptors are at golgi tendon

(D) Also called inverse myotatic reflex

97. Slow chronic pain sensation is carried via :

(A) A alpha fiber (B) A beta fiber

(C) A delta fiber (D) C fiber

98. Which one among the following is false about Neo-spino-thalamic tract :

(A) It is for fast pain

(B) Glutamate is the neurotransmitter

(C) Neurotransmitter is substance P

(D) Fast pain fibers terminate in Lamina I of dorsal horn

99. The muscles of the eye concern with the movements of the eye ball is called :

(A) Voluntary muscle (B) Extrinsic muscle

(C) Both (A) & (B) (D) Intrinsic muscle

100. Tear is produced by :

(A) Blind spot (B) Lacrimal gland

(C) Optic nerve (D) Corneal epithelium

101. Aqueous humor is secreted by :

(A) Iris (B) Ciliary process

(C) Retina (D) Cornea

102. Intraocular pressure is :

(A) Pressure due to aqueous humor (B) Pressure due to vitreous humor

(C) Both (A) & (B) (D) Pressure due to lacrimal secretion

103. Increased intraocular pressure leads to :

(A) Cataract (B) Miosis

(C) Horner's syndrome (D) Glaucoma

104. Pupillary dilatation is called :
(A) Mydriasis (B) Miosis
(C) Fogging (D) Glaucoma

105. What is the normal intra ocular pressure in human :
(A) 1-5 mm Hg (B) 10-20 mm Hg
(C) 25-35 mm Hg (D) 70-90 mm Hg

106. Astigmatism is a condition where transverse meridian curvature is less than vertical curvature, which can be cured by :
(A) Concave lens (B) Convex lens
(C) Cylindrical lens (D) Both (B) and (C)

107. Condition when lens become opaque and fail to transmit light is called :
(A) Cataract (B) Miosis
(C) Horner's syndrome (D) Glaucoma

108. In rhodopsin the retinal belongs to :
(A) Cis variety (B) Trans variety
(C) Both type (D) Either of these

109. The particular type of opsin that occurs in rhodopsin is :
(A) Retinene (B) Scotopsin
(C) Iodopsin (D) Photopsin

110. Which among the following is true about dark adaptation :
(A) When someone enters a dark room from brightly lit outdoor, first he is unable to anything but slowly begins to see more
(B) Poor visibility stimulates the photoreceptor cells
(C) As the person enters from bright light the threshold to stimulate photoreceptor cell is very high first but slowly the threshold decline
(D) All the above are true

111. What is false about Dichromats ?
(A) They have only two types of cone cells and the third type is missing
(B) They are mainly two types of Dichromats that are Protanopes and Deuteranopes
(C) Protanopes lack red sensitive cones
(D) Deuteranopes lack blue sensitive cones

112. A pouch like structure present in nose of some mammals and snake which contain receptors to identify food and exhibit sex behaviour is called :
(A) Olfactory filum (B) Vomeronasal organ
(C) Olfactory rod (D) Sustentacular cells

— Answer Sheet —

Q. No.	Ans	Q. No.	Ans	Q. No.	Ans	Q. No.	Ans	Q. No.	Ans	Q. No.	Ans
1.	**D**	21.	**A**	41.	**A**	61.	**A**	81.	**B**	101.	**B**
2.	**A**	22.	**A**	42.	**C**	62.	**D**	82.	**A**	102.	**A**
3.	**C**	23.	**B**	43.	**A**	63.	**C**	83.	**C**	103.	**D**
4.	**C**	24.	**D**	44.	**B**	64.	**B**	84.	**A**	104.	**A**
5.	**C**	25.	**C**	45.	**D**	65.	**C**	85.	**C**	105.	**B**
6.	**B**	26.	**A**	46.	**A**	66.	**C**	86.	**B**	106.	**C**
7.	**D**	27.	**B**	47.	**B**	67.	**B**	87.	**D**	107.	**A**
8.	**B**	28.	**A**	48.	**B**	68.	**C**	88.	**D**	108.	**A**
9.	**C**	29.	**B**	49.	**C**	69.	**D**	89.	**D**	109.	**B**
10.	**C**	30.	**C**	50.	**B**	70.	**A**	90.	**B**	110.	**D**
11.	**B**	31.	**C**	51.	**A**	71.	**D**	91.	**A**	111.	**D**
12.	**C**	32.	**B**	52.	**B**	72.	**C**	92.	**B**	112.	**B**
13.	**B**	33.	**C**	53.	**D**	73.	**C**	93.	**A**		
14.	**C**	34.	**C**	54.	**C**	74.	**B**	94.	**D**		
15.	**A**	35.	**B**	55.	**C**	75.	**A**	95.	**D**		
16.	**B**	36.	**B**	56.	**B**	76.	**C**	96.	**A**		
17.	**B**	37.	**C**	57.	**B**	77.	**A**	97.	**D**		
18.	**C**	38.	**A**	58.	**C**	78.	**B**	98.	**C**		
19.	**C**	39.	**C**	59.	**C**	79.	**A**	99.	**C**		
20.	**D**	40.	**C**	60.	**D**	80.	**B**	100.	**B**		